Medical Intelligence Unit

CTLA-4 in Autoimmune Disease

Flemming Pociot
Steno Diabetes Center
Denmark

CRC Press
Taylor & Francis Group
Boca Raton London New York

CRC Press is an imprint of the
Taylor & Francis Group, an **informa** business

CTLA-4 in Autoimmune Disease

Medical Intelligence Unit

CRC Press
Taylor & Francis Group
6000 Broken Sound Parkway NW, Suite 300
Boca Raton, FL 33487-2742

First issued in hardback 2019

ISBN-13: 978-1-58706-068-7 (hbk)
ISBN-13: 978-1-58706-127-1 (pbk)

Visit the Taylor & Francis Web site at
http://www.taylorandfrancis.com

and the CRC Press Web site at
http://www.crcpress.com

Library of Congress Cataloging-in-Publication Data

CIP applied for, but not received at time of publication.

CONTENTS

EDITOR

Flemming Pociot
Steno Diabetes Center
Denmark
Chapter 1

CONTRIBUTORS

Maria-Luisa Alegre
University of Chicago
Department of Medicine
Hematology and Oncology Section
Chicago, Illinois, U.S.A.
Chapter 2

Klaus Badenhoop
Medizinische Klinki I
Klinikum der
Johann Wollfgang Goethe-Universitat
Frankfurt am Main, Germany
Chapter 8

Raffaella Buzzetti
Dipartimento di Scienze Cliniche
Universita degli studi di Roma
"La Sapienza"
Roma, Italia
Chapter 9

Isabella Cascino
Istuto di Biologia Cellulare
CNR, Campus Buzzati-Traverso
Monterotondo (Roma), Italy
Chapter 9

David I. Daikh
University of California
San Francisco and San Francisco VA
Medical Center
Arthritis/Immunology Section
San Francisco, California, U.S.A.
Chapter 3

Thomas F. Gajewski
University of Chicago
Department of Medicine
Hematology and Oncology Section
Chicago, Illinois, U.S.A.
Chapter 2

Rebecca J. Greenwald
Brigham and Women's Hospital
and Harvard Medical School
Department of Pathology
Boston, Massachusetts, U.S.A.
Chapter 7

E. Helen Kemp
Division of Clinical Science
Northern General Hospital
University of Sheffield
Sheffield, U.K.
Chapter 5

Fadi G. Lakkis
Department of Medicine
Section of Nephrology
Yale University School of Medicine
New Haven, Connecticut, U.S.A.
Chapter 10

Yvette Latchman
Brigham and Women's Hospital
and Harvard Medical School
Department of Pathology
Boston, Massachusetts, U.S.A.
Chapter 7

Ann Kari Lefvert
Karolinska Hospital
Immunological Research Unit
Center for Molecular Medicine
Stockholm, Sweden
Chapter 6

Stamatis-Nick Liossis
Research Immunology Laboratory
First Department of Propedeutic Medicine
Athens University Medical School
Athens, Greece
Chapter 4

Lorenza Nisticò
Isttuto di Biologia Cellulare
CNR, Campus Buzzati-Traverso
Monterotondo (Roma), Italy
Chapter 9

Paolo Pozzilli
Universita di Roma
Campus Biomedico
Roma, Italia
Chapter 9

David M. Rothstein
Department of Medicine
Section of Nephrology
Yale University School of Medicine
New Haven, Connecticut, U.S.A.
Chapter 10

Peter P. Sfikakis
Research Immunology Laboratory
First Department of Propedeutic Medicine
Athens University Medical School
Athens, Greece
Chapter 4

Arlene H. Sharpe
Brigham and Women's Hospital
and Harvard Medical School
Department of Pathology
Boston, Massachusetts, U.S.A.
Chapter 7

Elizabeth A. Waterman
Division of Clinical Science
Northern General Hospital
University of Sheffield
Sheffield, U.K.
Chapter 5

Anthony P. Weetman
Division of Clinical Science
Northern General Hospital
University of Sheffield
Sheffield, U.K.
Chapter 5

David Wofsy
Rheumatology Section
San Francisco VA Medical Center
San Francisco, California, U.S.A.
Chapter 3

PREFACE

There have been remarkable advances in our understanding of the molecular pathogenesis of autoimmunity over the last decade. It is becoming increasingly clear that different autoimmune diseases share common features, some of which may indeed be critical for the very initiation of the autoimmune process.

The Human Genome Project is generating an amount of data unprecedented in biology. Science has never before had the tools to characterize the pathophysiological nuances and inherited variations that interact over time and lead to common diseases such as autoimmune diseases. It is thus, expected that our knowledge about biology will increase sharply over the next decade.

In almost all autoimmune diseases the most important genetic factor is the MHC locus, and in some cases it is the only locus for which statistically significant association has yet been found. The only other 'major' player identified in several autoimmune diseases is the CTLA-4 (cytotoxic T-lymphocyte associated antigen 4) locus—a molecule involved in the costimulatory signaling regulating T-cell function. Evidence for CTLA-4 comes from both functional and genetic studies.

In this book, the intention has been to provide the reader, in a single volume, with an up-to-date summary of the data indicating a role of CTLA-4 in autoimmunity. Authoritative international investigators have written the chapters.

Although it was attempted to cover all aspects of autoimmunity in which CTLA-4 have been implicated, it is realized that some areas may not be detailed in this volume. Nevertheless, this volume is one of the first wide-ranging attempts to conceive the role of molecules outside the MHC region as a common denominator for autoimmune diseases.

The perspectives of this field are immense, not only for understanding the pathology of the diseases but also for providing new treatment modalities. Notably, interferon β for multiple sclerosis and tumor necrosis factor α antagonists for rheumatoid arthritis and Crohn's disease are the first new treatments for autoimmunity approved by the Food and Drug Administration in 20 years. CTLA4-immunoglobulins or other molecules manipulating the costimulatory signals regulating T-cell function may well be next on this list.

Last but not least, I am indebted to the many authors around the world who have so generously devoted their knowledge, energy and time to the creation of this book.

Flemming Pociot
Steno Diabetes Center

CHAPTER 1

Autoimune Disorders—A Common Link?

Flemming Pociot

The immune system has evolved to protect multicellular organisms from pathogens. It is therefore perplexing that this system turns on the individual, in some cases precipitating catastrophic autoimmune disease. The capability of the immune system to destroy a variety of cell types is evident, both from the human diseases themselves and their excellent animal models. The tissue attacked varies from the beta cells of the endocrine pancreas in Type 1 (insulin-dependent) diabetes mellitus (T1DM), joint components in rheumatoid arthritis (RA), the myelin of the central nervous system in multiple sclerosis (MS) and major structures of the cell nucleus in systemic lupus erythematosis (SLE). In addition, it seems likely that a substantial amount of human pathology will ultimately be explained by damage that is mediated by the immune system in common diseases such as atherosclerosis. Thus, the understanding of the underlying mechanisms of autoimmunity is a key goal for the improvement of human health.

Autoimmune diseases affect some 5-7% of adults in North America and Europe, Table 1. They have major socio-economic impact, as they are associated with debilitation during the most productive years of life.

Despite the avalanche of molecular and immunogenetic data and improvements in seriologic diagnosis, the specific etiologies of virtually all human autoimmune diseases are unknown. For some autoimmune diseases, there may be a very long latency period between disease onset and clinical presentation. Existing therapies tend only to be partially successful and often associated with a variety of serious side effects. However, even in the absence of a complete understanding of the underlying mechanisms that confer susceptibility to autoimmune diseases, it may be possible to devise successful therapies by interfering with one or more of the specific pathways characteristic of tissue destruction.

The Immune System in Autoimmunity

Autoimmune diseases occur when an immune response is directed against a specific organ or a number of organs and systems within the individual. For an autoimmune disease to occur, a repertoire of self-reactive lymphocytes must exist. This repertoire must become activated and effector cells must deliver pathologic damage to a given self-tissue. Since autoimmune diseases only occur in a small subset of the total population, it is believed that regulatory mechanisms prevent this in the majority of cases. Thus, these regulatory mechanisms must fail in those with autoimmune disease.

The major mechanism of T-cell tolerance is the deletion of self-reactive T cells in the thymus. This induction of central tolerance requires the presence of autoantigens in the thymus.[1-4] Not all self-antigens occur in the thymus, which necessitates the existence of

CTLA-4 in Autoimmune Disease, edited by Flemming Pociot. ©2004 Eurekah.com.

Table 1. Population frequencies and familial clustering of most common autoimmune diseases

Disease	Population Frequency (%)	Sibling Risk (%)	λ_s	HLA λ_s	HLA Contribution to Familial Clustering*
Psoriasis	2.5	17	7	ND	
Rheumatoid arthritis	1.0	8	8	1.6	22%
Grave's disease	0.5	7.5	15	ND	
Type 1 diabetes	0.4	6	15	2.5	34%
Multiple sclerosis	0.1	2	20	2.4	29%
Ulcerative colitis	0.1	1.2	12	8.3	85%
Systemic lupus erythematosus	0.1	2	20	ND	
Celiac disease	0.05	3	60	5.2	40%

Population prevalences and sibling risk values are averaged values based on published data and mainly adapted from Vyse and Todd.[23] The values are gross estimates since they have wide confidence intervals.
*: The calculation of these values is assuming a simple multiplicative model of inheritance. ND: not determined.

peripheral mechanisms to participate in T-cell tolerance. Attenuation of peripheral mechanisms of self-tolerance may result in the development of autoimmune diseases.[5]

The peripheral mechanisms involved in the induction of tolerance include immunologic ignorance, deletion, anergy, inhibition and suppression[6] (Fig. 1).

The immune response to foreign antigens is of a dynamic nature and a series of discrete events are involved in peripheral tolerance.[7,8] In this dynamic process, there is obviously a potential for multiple costimulatory signals that may be delivered by different cell types at distinct stages of the immune response. The engagement of the T-cell receptor with peptide presented in the context of major histocompatibility complex (MHC) class II molecules is the first step. CD28 engagement by B7-1 (CD80) and B7-2 (CD86) on resting T helper cells provides a critical signal for initial cell cycle progression, interleukin-2 production and clonal expansion. Cytotoxic T lymphocyte antigen 4 (CTLA-4, CD152) belongs to the same family of costimulatory molecules as CD28. In contrast to CD28, CTLA-4 is induced upon activation and delivers a negative signal to the activated T cell, opposing CD28-mediated costimulation[9-11] (for details see chapter 2).

Autoimmune Diseases

Autoimmune diseases are divided broadly into two types: organ specific and non-organ specific or systemic (Table 2). In practice, a continuum of the diseases exists. The classification is based not so much on the location of the disease, as the target site for the immune response. Thus, in organ specific diseases the response is targeted against a single organ or tissue while in non-organ specific diseases the target antigen is found systematically throughout the body.

It seems almost certain that many (most?) individuals with an autoimmune repertoire never develop autoimmune disease. Stochastic environmental factors thus play an important role in the precipitation of these diseases, and it has been suggested that the status of this repertoire can be depicted as a phase of dynamic instability.[12] Further elaboration of

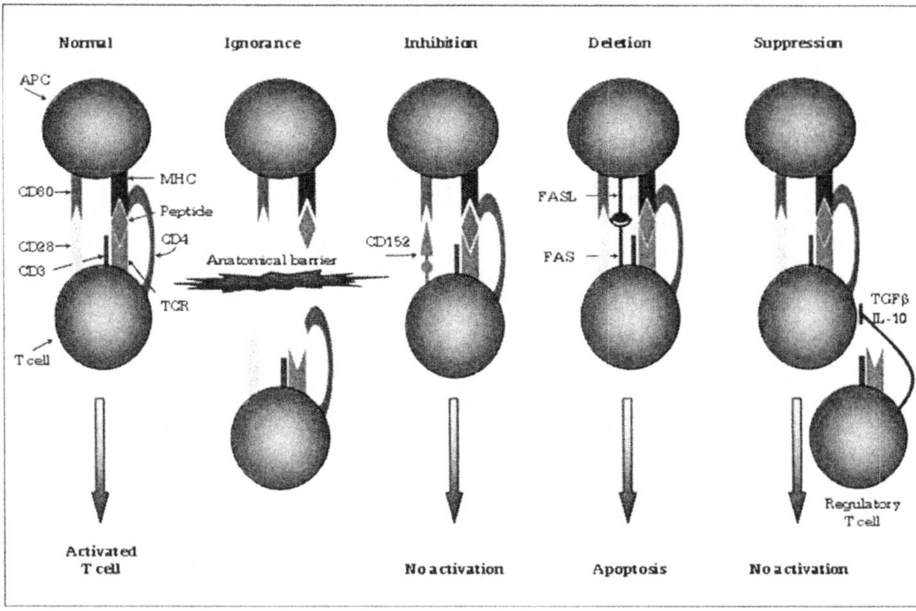

Figure 1. Peripheral mechanisms involved in induction of tolerance. The molecular components involved in normal activation of T cells are outlined in the left panel. T cells which are separated – e.g., by the blood-brain barrier – from their specific antigen cannot become activated, a situation referred to as immunological ignorance. CTLA-4 (CD152) inhibits the activation of T cells by binding to CD80 (B7-1) with higher affinity than the costimulatory receptor CD28. T cells that express FAS (CD95) can be induced to undergo apoptosis by interacting with cells expressing FAS ligand (FASL). This process is known as deletion. Regulatory T cells can also suppress other T cells through the production of inhibitory cytokines such as transforming growth factor β (TGFβ) and interleukin 10 (IL-10). APC: antigen presenting cell; MHC: major histocompatibility complex; TCR: T-cell receptor. (Adapted from ref. 6).

this concept allows mathematical formulation of the pathogenetic models for autoimmune diseases, which eventually may lead to better characterization of the autoimmune processes and provide the rationale for the design of new intervention modalities.

Otherwise, very little is known about environmental factors precipitating autoimmune diseases. It has been hypothesized that infections with acute or persistent DNA or RNA viruses induce, accelerate or enhance autoimmune responses and cause autoimmune disease.[13-15]

The understanding of selection and activation processes of autoreactive T cells is of utmost importance for the understanding of the nature of the autoantigen. In some cases the key antigens have been identified (Table 3). Self-antigens have been reported in most autoimmune diseases including Type 1 diabetes, multiple sclerosis, autoimmune thyroid disease, and rheumatoid arthritis. Interestingly, not all key antigens have to be encoded by the human genome. It has been shown that antigens playing an important role in animal models of human inflammatory bowel diseases include antigens derived from gut bacteria.[16-18]

When autoantibodies or autoreactive cells are found in an individual with detectable pathology, there may be several possible explanations for this observation. Firstly, the autoantibodies or autoreactive cells may have caused the lesion; secondly, some other event may have caused the lesion, which then led to an autoimmune response, or, finally, one particular factor may have caused both the lesions and the autoimmune response. It is

Table 2. Classification of autoimmune diseases

Organ-specific autoimmune diseases	
	Type 1 diabetes mellitus
	Thyroid autoimmune disorders (Grave's and Hashimoto's diseases)
	Celiac disease
	Addison's disease
	Multiple sclerosis
	Myastenia gravis
Non-organ specific (systemic) autoimmune diseases	
	Rheumatoid arthritis
	Psoriasis
	Inflammatory bowel diseases
	Systemic lupus erythematosus

The spectrum of autoimmune diseases ranges from organ-specific diseases to non-organ specific diseases. Some of the diseases like multiple sclerosis, myasthenia gravis and rheumatoid arthritis display features of both types.

likely that different pathogenetic processes will lead to inseparable phenotypes for most of the autoimmune diseases. Thus, autoimmune diseases may be a result of an aberrant immune response and/or of increased vulnerability of the target organ/cells. Which comes first in each case may be impossible to distinguish.

Genetics of Autoimmune Diseases

There is little doubt that genetic factors play a critical role in predisposition to autoimmune diseases. Autoimmune diseases do not follow a simple Mendelian pattern of inheritance. Multiple genes that act in concert to cause disease may influence the phenotype. Such a genetic etiology is called polygenic, or complex inheritance. Multifactorial inheritance is an extension of polygenic inheritance, where additional, nongenetic (environmental) factors may also be involved. Hence, autoimmune diseases are classified as multifactorial diseases although it is not precisely known how many genetic loci are involved or how they interact with environmental factors. These polygenes are referred to as susceptibility genes, which only act to modify the risk for disease. As individual susceptibility genes they are neither necessary nor sufficient for development of the disease, i.e., some gene variant carriers may never develop the disease while some non-carriers may have the disease. This lack of definite correlation between genotype and phenotype is a major complicating factor in mapping and cloning genes for autoimmune diseases and other common diseases.

Genetic factors, environmental factors or both cause clustering of disease in families. The degree of familial clustering can be estimated by the increased risks in siblings over population prevalence (λ_s).[19] If this ratio, λ_s, is close to 1.0, then there is no evidence for familial clustering. The λ_s-value for some autoimmune diseases is listed in Table 1. Note that for all the listed diseases the λ_s value greatly exceeds 1.0. Although, the λ_s-values are sufficiently high to support a role for genetic factors, they are, however, lower than λ_s-values of monogenic disorders caused by rare highly penetrant gene mutations. In general, the lower the λ_s, the more difficult it will be to identify, i.e., clone, the genes involved. It will also be difficult to prove which of the possibly many polymorphisms in a disease-associated chromosomal segment are of etiological importance. Familial clustering of autoimmune

Table 3. Identified key antigens in autoimmune diseases

Disease	Antigen(s)
SLE	Histones
RA	Collagen II, aggrecan
Graves' disease	TSHr (stimulating)
Hashimoto's disease	Thyroid peroxidase, thyroglobulin, TSHr (blocking)
Myastenia gravis	Nicotine acetylcholine receptor
Multiple sclerosis	Myelin basic protein (MBP)
Addison's disease	21-OH
T1DM	Insulin, GAD, IA-2
Celiac disease	Antigliadin, antiendomysial, transglutaminase
Vitiligo	Tyrosinase/tyrosinase-related proteins

Some key antigens recognized in different autoimmune diseases. For most of the diseases several additional antigens have been identified. For more details see the disease-specific chapter of this volume.

diseases may also be due to shared environmental factors, e.g., viral infections and certain nutritional factors.

Two main approaches have been used to dissect the genetics of autoimmune diseases—candidate-gene analysis, and whole-genome screening approaches. Candidate genes have been derived either from obvious physiological/biochemical information or from the fact that the gene is expressed in affected tissue, e.g., pancreatic islets, synovial components, thyroid cells, or has a function known to be affected in the disorder. After pioneering work in Type 1 diabetes,[20,21] it seems likely that most autoimmune diseases are polygenic.[22]

The occurrence of common features of autoimmune diseases and the co-occurrence of multiple autoimmune diseases in the same individual or family support the notion that there may be common genetic factors that predispose to autoimmunity. This is indeed the case also for autoimmune diseases in mouse.[23] Recently, the existence of overlapping susceptibility loci between different human autoimmune diseases has been demonstrated.[24-27]

This clustering of autoimmune susceptibility loci suggests that there may be related genetic factors contributing to susceptibility of clinically distinct diseases, and that the genes to be identified in these clusters are most likely involved in primary or secondary regulation of the immune system.

Do We Know the Genes?

Genes located within the HLA region are by far the most important in conferring susceptibility to most autoimmune diseases, but several other susceptibility regions have been implicated.

The Human Leukocyte Antigen (HLA) System

HLA class II molecules are assembled in the endoplasmatic reticulum and are transported through the Golgi apparatus to the cytosol where they take up degraded antigen. They are then transported to the cell surface to present antigen to $CD4^+$ T cells. This interaction between HLA class II molecules and T cells makes the HLA class II region a strong candidate region for diseases, which develop as a result T-cell mediated autoimmunity.

Genes of the HLA region on chromosome 6p21 have been implicated in most autoimmune diseases. The HLA class II region accounts for 20-85% of familial clustering of different autoimmune diseases (Table 1). In some cases aberrant expression of HLA class II molecules has been demonstrated on target cells and/or on activated lymphocytes in patients with autoimmune disease. However, in most cases the specifically pathogenetic mechanism(s) underlying the observed HLA-disease association is not fully understood.

Other Genes

One locus at chromosome 2q33 has been associated with multiple autoimmune diseases, which suggests that it encodes a general susceptibility gene for autoimmunity.[26] Candidate genes at this locus include those that encode the T-cell costimulatory receptors, CD28 and CTLA-4. Although the evidence linking either of these molecules is inconclusive, CTLA-4 is the leading candidate molecule. This is partly due to the fact that markers used in linkage and association studies map within the gene encoding CTLA-4 as detailed in the other chapters of this volume. In addition, knowledge of the functional properties of CTLA-4 supports this (detailed in chapter 2). However, recent data from the completion of the human genome sequence and from a series of functional studies using genetic knockouts suggest that an additional costimulatory receptor, structurally and functionally related to CD28 and CTLA-4, the inducible costimulator (ICOS),[28-29] may be an additional candidate for this region on chromosome 2q33. Sequence data closely link ICOS to CD28 and CTLA-4. The gene encoding ICOS is separated only by approximately 100.000 bases from that encoding CTLA-4 (www.genome.ucsc.edu) indicating that both genes are well within the limits of resolution of current genetic studies. Studies using ICOS knockouts[30-32] have demonstrated that ICOS is essential for development of normal T cell help and plays a protective role during induction of experimental autoimmune disease.[30] This suggests that there may be more autoimmune disease genes at chromosome 2q33 and that the immune response is a dynamic process where there is obviously a potential for multiple costimulatory signals.

A long list of other genes has been evaluated for the different autoimmune diseases. Some of these are mentioned in the following chapters, but except for HLA-region genes and CTLA-4, no other genes have convincingly been shown to influence susceptibility to several autoimmune diseases.

Immunomodulation as Treatment of Autoimmune Diseases

Manipulation of the interaction between the B7 family of molecules and their ligands has been envisioned as a potential strategy for achieving therapeutically useful immunosuppression or tolerance. CTLA4-Ig binds B7 molecules with higher affinity than does CD28 and therefore acts as a potent competitive inhibitor of B7-CD28-mediated T-cell costimulation. Blockade of the CD28 pathway by CTLA4-Ig may permit the development of immunological tolerance by exploiting normal mechanisms of tolerance to self-antigens. Manipulation of costimulatory signals does not require identification of the antigen or of the individual polymorphic HLA and T-cell receptor genes/molecules. This is particularly advantageous for therapy of autoimmune diseases, where the precise antigen eliciting T-cell activation may not be known.

Recently, CTLA4-Ig has entered Phase I clinical trails for the treatment of psoriasis,[33-34] a T-cell mediated skin diseases, and treatment of graft-versus-host disease in allogenic bone marrow transplantation.[35] Large clinical randomized trials on the use of

CTLA4-Ig are missing. Nevertheless, its immunosuppressive effect coupled with features, such as specificity of interaction and low toxicity, make CTLA4-Ig a promising new therapeutic agent for induction of donor-specific immunological tolerance, the ultimate goal of clinical immunosuppression.

Very interestingly, effects of CTLA4-Ig treatment have been demonstrated in mouse models of systemic lupus erythematosus (SLE), even when the treatment was delayed until the most advanced stage of the diseases[36] (see chapter 3 for more details). The effectiveness of CTLA4-Ig on established disease is particularly interesting, since most autoimmune diseases are diagnosed after initial responses to autoantigens. Also the effect of CTLA4-Ig in experimental models of RA is promising (see chapter 4 for details).

Consistent with rodent studies, CTLA4-Ig is well-tolerated in non-human primate models.[37] In man, in Phase I clinical trials of psoriasis, a four-dose i.v. regimen of CTLA4-Ig, up to 50 mg/kg, was well tolerated and did not cause any immediate hypersensitivity reactions, or decrease serum Ig levels and did not lead to anti-CTLA4-Ig antibody production.[33-34]

Thus, manipulation of the costimulatory signals regulating T-cell function, e.g., by CTLA4-Ig, seems to be a very promising treatment modality in autoimmunity.

References

1. Pugliese A, Zeller M, Fernandez A et al. The insulin gene is transcribed in the human thymus and transcription levels correlate with allelic variation at the *INS VNTR-IDDM2* susceptibility locus for type-1 diabetes. Nat Genet 1997; 15:293-297.
2. Vafiadis P, Bennett S, Todd J et al. Insulin expression in human thymus is modulated by *INS VNTR*. Nat Genet 1997; 15:289-292.
3. Akkaraju S, Ho WY, Leong D et al. A range of cd4 t-cell tolerance—partial inactivation to organ-specific antigen allows nondestructive thyroiditis or insulitis. Immunity 1997; 7:255-271.
4. Klein L, Klugmann M, Nave KA et al. Shaping of the autoreactive T-cell repertoire by a splice variant of self protein expressed in thymic epithelial cells. Nat Med 2000; 6:56-61.
5. Bluestone JA. Is ctla-4 a master switch for peripheral t-cell tolerance? J Immunol 1997; 158:1989-1993.
6. Kamradt T, Mitchison NA. Advances in immunology: Tolerance and autoimmunity. N Engl J Med 2001; 344:655-664.
7. Garside P, Ingulli E, Merica RR et al. Visualization of specific b-lymphocyte and t-lymphocyte interactions in the lymph-node. Science 1998; 281:96-99.
8. Coyle AJ, Gutierrez-Ramos JC. The expanding B7 superfamily: Increasing complexity in costimulatory signals regulating T cell function. Nat Immunol 2001; 2:203-209.
9. Lindsten T, Lee KP, Harris ES et al. Characterization of ctla-4 structure and expression on human t-cells. J Immunol 1993; 151:3489-3499.
10. Linsley PS, Greene JL, Brady W et al. Human b7-1 (cd80) and b7-2 (cd86) bind with similar avidities but distinct kinetics to cd28 and ctla-4 receptors. Immunity 1994; 1:793-801.
11. Walunas TL, Lenschow DJ, Bakker CY et al. Ctla-4 can function as a negative regulator of T-cell activation. Immunity 1994; 1:405-413.
12. Freiesleben-De Blasio B, Bak P, Pociot F et al. Onset of type 1 diabetes: a dynamical instability. Diabetes 1999; 48:1677-1685.
13. von Herrath MG, Oldstone MBA. Virus-induced autoimmune-disease. Curr Opin Immunol 1996; 8:878-855.
14. Oldstone MBA. Viruses and autoimmune-diseases. Scand J Immunol 1997; 46:320-325.
15. Jun HS, Yoon-JW. The role of viruses in Type I diabetes: two distinct cellular and molecular pathogenic mechanisms of virus-induced diabetes in animals. Diabetologia 2001; 44:271-285.
16. Mizoguchi A, Mizoguchi E, Tonegawa S et al. Alteration of a polyclonal to an oligoclonal immune-response to cecal aerobic bacterial-antigens in tcr-alpha mutant mice with inflammatory bowel-disease. Int Immunol 1996; 8:1387-1394.
17. Brandwein SL, Mccabe RP, Cong YZ et al. Spontaneously colitic c3h/hejbir mice demonstrate selective antibody reactivity to antigens of the enteric bacterial- flora. J Immunol 1997; 159:44-52.

18. Cong YZ, Brandwein SL, Mccabe RP et al. Cd4(+) t-cells reactive to enteric bacterial-antigens in spontaneously colitic c3h/hejbir mice—increased t-helper cell-type-1 response and ability to transfer disease. J Exp Med 1998; 187:855-864.
19. Risch N. Assessing the role of HLA-linked and unlinked determinants of disease. Am J Hum Genet 1987; 40:1-14.
20. Davies JL, Kawaguchi Y, Bennett ST et al. A genome-wide search for human susceptibility genes. Nature 1994; 371:130-136.
21. Hashimoto L, Habita C, Beressi J et al. Genetic mapping of a susceptibility locus for insulin-dependent mellitus on chromosome 11q. Nature 1994; 371:161-164.
22. Griffiths MM, Encinas JA, Remmers EF et al. Mapping autoimmunity genes. Curr Opin Immunol 1999; 11:689-700.
23. Vyse TJ, Todd JA. Genetic analysis of autoimmune disease. Cell 1996; 85:311-318.
24. Becker K, Simon R, Bailey-Wilson J et al. Clustering of non-major histocompatibility complex susceptibility candidate loci in human autoimmune diseases. Proc Natl Acad Sci USA 1998; 95:9979-9984.
25. Becker KG. Comparative genetics of type 1 diabetes and autoimmune disease—Common loci, common pathways? Diabetes 1999; 48:1353-1358.
26. Kristiansen OP, Larsen ZM, Pociot F. CTLA-4 in autoimmune diseases—a general susceptibility gene to autoimmunity? Genes Immun 2000; 1:170-184.
27. Merriman TR, Cordell HJ, Eaves IA et al. Suggestive evidence for association of human chromosome 18q12-q21 and its orthologue on rat and mouse chromosome 18 with several autoimmune diseases. Diabetes 2001; 50:184-194.
28. Hutloff A, Dittrich AM, Beier KC et al. ICOS is an inducible T-cell co-stimulator structurally and functionally related to CD28. Nature 1999; 397:263-266.
29. Coyle AJ, Lehar S, Lloyd C et al. The CD28-related molecule ICOS is required for effective T cell-dependent immune responses. Immunity 2000; 13:95-105.
30. Dong C, Juedes AE, Temann UA et al. ICOS co-stimulatory receptor is essential for T-cell activation and function. Nature 2001; 409:97-101.
31. Tafuri A, Shahinian A, Bladt F et al. ICOS is essential for effective T-helper-cell responses. Nature 2001; 409:105-109.
32. McAdam AJ, Greenwald RJ, Levin MA et al. ICOS is critical for CD40-mediated antibody class switching. Nature 2001; 409:102-105.
33. Abrams JR, Lebwohl MG, Guzzo CA et al. CTLA4Ig-mediated blockade of T-cell costimulation in patients with psoriasis vulgaris. J Clin Invest 1999; 103:1243-1252.
34. Abrams JR, Kelley SL, Hayes E et al. Blockade of T lymphocyte costimulation with cytotoxic T lymphocyte-associated antigen 4-immunoglobulin (CTLA4Ig) reverses the cellular pathology of psoriatic plaques, including the activation of keratinocytes, dendritic cells, and endothelial cells. J Exp Med 2000; 192:681-693.
35. Guinan EC, Boussiotis VA, Neuberg D et al. Transplantation of anergic histoincompatible bone marrow allografts. N Engl J Med 1999; 340:1704-1714.
36. Finck BK, Linsley PS, Wofsy D. Treatment of murine lupus with ctla4ig. Science 1994; 265:1225-1227.
37. Kirk AD, Harlan DM, Armstrong NN et al. Ctla4-ig and anti-cd4o ligand prevent renal-allograft rejection in primates. Proc Natl Acad Sci (USA) 1997; 94:8789-8794.

Chapter 2

CTLA-4: Its Role in the Immune Response

Maria-Luisa Alegre and Thomas F. Gajewski

Introduction

T lymphocytes are essential for host defense against many viral or parasitic infections, and also contribute to defense against tumors. In addition, T cells mediate rejection of transplanted organs, and, if inappropriately activated to recognize self-antigens, can cause autoimmune diseases. Under normal conditions, the magnitude of a T cell response rises and falls in a predictable fashion and the processes of both activation and quiescence of T lymphocytes are carefully regulated. T cell activation depends on recognition by the T cell receptor (TCR) of specific antigenic peptide in the context of major histocompatibility complex (MHC) molecules expressed by antigen presenting cells (APCs) such as dendritic cells, B cells, or macrophages. Additional signals delivered by costimulatory receptors such as CD28 or tumor necrosis factor receptor (TNFR) family members are also required for complete T cell activation and differentiation to occur. Following T cell activation, inhibitory receptors such as CTLA-4 or the more recently described PD-1[1] become expressed, and can promote the termination of an adaptive immune response. This chapter will focus on the regulation and function of CTLA-4.

CTLA-4 Characteristics and Properties

CTLA-4 (CD152) is a 30-40 kDa heavily glycosylated protein identified in 1987 by Brunet and colleagues as a result of a search for cDNAs preferentially expressed by cytotoxic T lymphocytes (CTL).[2] The CTLA-4 gene has been mapped to mouse chromosome 1 and human chromosome 2.[3,4] CTLA-4, like CD28, belongs to the immunoglobulin superfamily, and shares around 50% homology with CD28. It is composed of a V-like extracellular domain, an intracellular domain, and a short cytoplasmic tail of 36 amino acids, with no predicted intrinsic enzymatic activity. Similar to CD28, CTLA-4 can bind B7-family members on APCs, namely B7-1 (CD80) and B7-2 (CD86), although the affinity for these ligands is approximately 20-fold higher than that of CD28.[5] Flow cytometry analysis performed with a soluble form of CTLA-4 on APCs from mice deficient in both CD80 and CD86 has suggested that these are the only ligands existing for CTLA-4.[6] CTLA-4 is expressed as a disulfide-linked homodimer on activated T cells and dimerization appears required for the formation of high-avidity complexes with B7 ligands and for transmission of signals.[7,8] Recently, the crystal structure of CTLA-4 has been solved. CTLA-4 has an unusual mode of dimerization that places the B7 binding sites distal to the dimerization interface.[9] This organization allows each CTLA-4 dimer to bind two bivalent B7 molecules. The extracellular binding sites for CD80 and CD86 also have been mapped.[10-12]

CD80 is expressed on activated B cells and monocytes, as well as on dendritic cells and Langerhans cells. CD80 is upregulated on B cells upon MHC class II engagement, and its expression is further enhanced by IL-2 and IL-4.[13,14] CD86 is 25% homologous with CD80, and is expressed on resting monocytes, dendritic cells and T cells, and on activated NK cells, B cells, and other APCs. Both CD80 and CD86 can be expressed on activated T cells, but the role of these molecules on T cells has remained elusive. Data obtained with human T cells indicate that CD86 is expressed as a hypoglycosylated form, has marked reduced affinity for CTLA-4, and does not bind CD28, suggesting that B7 family members expressed on T cells have minimal functional activity.[15] Whether CTLA-4 preferentially binds to CD80 or CD86 in vivo has been a matter of debate. It has been suggested that because CD86 is either constitutively expressed or rapidly induced it may preferentially bind CD28 whereas later kinetic expression of CTLA-4 and CD80 may favor this association. Although some experiments are consistent with this hypothesis,[16,17] no definitive evidence is yet available.

In addition to T lymphocytes, expression of CTLA-4 at the protein level has also been reported on activated B cells, either in the presence or absence of T cells.[18,19] It appears that crosslinking of CTLA-4 during B cell receptor stimulation can result in reduced IL-4-driven isotype switching, with decreased levels of IgG1 and IgE, a result that has relevance for allergy and autoimmune diseases.

In addition to the membrane-bound form of CTLA-4, a soluble secreted form has recently been described.[20,21] Messenger RNA (mRNA) for this shorter form occurs through alternative splicing, is found preferentially in resting T and B cells, and its transcription is reduced upon T cell activation. A low concentration of soluble CTLA-4 molecules can be found in the serum of some healthy volunteers but higher levels have been observed in patients with autoimmune thyroid disease[22] The overall significance of these observations is not yet clear.

Regulation and Localization of CTLA-4 Expression

Unlike CD28, CTLA-4 is not constitutively expressed on resting T cells. Upon TCR stimulation, CTLA-4 expression is induced, peaks 48-72h following stimulation, and progressively decreases to very low levels[23] (Fig. 1). Expression of CTLA-4 is further upregulated by signals delivered by the CD28 and IL-2 receptors .[24] In particular, CD28 ligation has been shown to increase both CTLA-4 transcription and mRNA stabilization (25, 26). In addition, expression of CTLA-4 requires cell division, as drugs that induce cell-cycle arrest can effectively prevent its upregulation. Surprisingly, most of CTLA-4 molecules are found in intracytoplasmic pools with only about 10% being expressed at the cell surface at any given time. The intracellular organelles containing CTLA-4 molecules include the Golgi apparatus, endocytic vesicles, and lysosomes. CTLA-4 molecules traffic very actively from intracellular localizations to the cell surface before being rapidly endocytosed into clathrin-coated vesicles .[27] It is hypothesized that limiting the amount of CTLA-4 expressed on the surface regulates its function in the T cell by restricting ligation to B7 molecules and therefore also regulates the outcome of immune responses.

The targeting of CTLA-4 to clathrin-coated vesicles is mediated by binding of the cytoplasmic tail of CTLA-4 to AP50, the medium chain subunit of the clathrin-associated adaptor complex AP-2.[28-31] The interaction requires the presence of the unphosphorylated

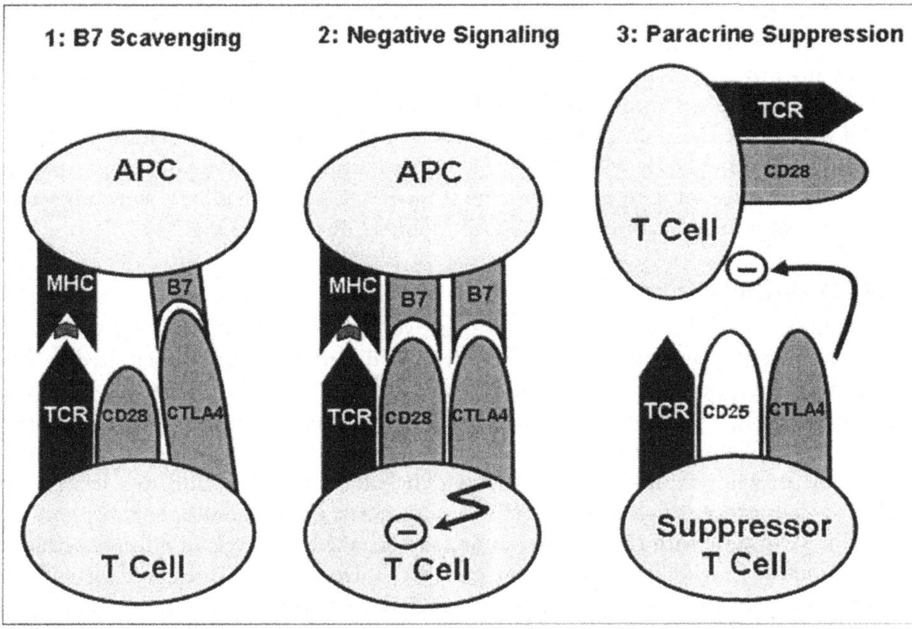

Figure1. CTLA-4 is induced to be expressed on T cells upon ligation of the TCR and of costimulatory receptors.

peptide sequence 199-GVYVKM-204 in the cytoplasmic tail of CTLA-4. Mutation or phosphorylation of amino acid residue Y201 abrogates the interaction with AP50, resulting in the accumulation of CTLA-4 at the cell surface. A current model proposes that, upon receptor ligation to B7, the cytoplasmic tail of CTLA-4 becomes phosphorylated, thus decreasing binding to AP-50 and preventing endocytosis while promoting interactions with signaling enzymes that mediate its inhibitory effect. A recent report indicated that internalized CTLA-4 is rapidly degraded within lysosomes, and that the degradation process may also serve as one of the mechanisms regulating CTLA-4 expression.[32] Upon TCR stimulation, some lysosomes containing internalized CTLA-4 molecules become secreted, in parallel with upregulation of surface CTLA-4, suggesting that a fraction of CTLA-4 is recycled to the plasma membrane following endocytosis.

Function of CTLA-4 in vitro

Because of the homology with CD28, and because soluble anti-CTLA-4 mAbs induced augmentation of T cell responses in vitro in the presence of APCs, it was initially hypothesized that CTLA-4 was a costimulatory molecule for T cell activation. However, it was soon clear that the anti-CTLA-4 antibodies used were all blocking ligation of CTLA-4 rather than mimicking binding by B7 ligands. Therefore, the augmentation of T cell responses appeared to be the consequence of interfering with the delivery of a negative signal. Indeed, the vast majority of reports to date indicate that the predominant function of CTLA-4 is inhibitory to activated T cells.[23,33-35] Although it has been suggested that, under particular circumstances, CTLA-4 cross-linking may promote T cell death,[36,37] this is controversial and may be secondary to the reduced production of growth factors. Indeed, multiple studies have indicated that cross-linking of CTLA-4 leads to reduced production of IL-2 via transcriptional inhibition, decreased expression of IL-2R, and cell-cycle arrest. The latter correlates with inhibition of expression of cyclin D3 and of the cyclin-dependent kinases cdk4 and cdk6.[38]

In order to exert its inhibitory effects, ligation of CTLA-4 needs to be provided by the same cell that presents antigen to the TCR.[39] Thus, following interaction with cells transfected with single chain mAbs specific for CD3 or for CTLA-4, proliferation of activated T cells was inhibited only when anti-CTLA-4 was expressed by the same cells expressing anti-CD3. Consistent with this requirement, surface CTLA-4 has been shown by confocal microscopy to co-localize with the TCR upon TCR cross-linking.[40]

CTLA-4-Deficient Mice

Consistent with CTLA-4 having a negative regulatory effect on T cells, CTLA-4-deficient mice display a major lymphoproliferative disorder. This is characterized by T cell accumulation in peripheral lymphoid tissues as well as T cell infiltration of most main organs, resulting in animal wasting and death by 3-5 weeks of age.[41,42] Similarly, chronic blockade of CTLA-4 from birth has recently been shown to result in an autoimmune syndrome in wildtype mice.[43] In CTLA-4-deficient mice, $CD4^+$ and to a lesser extent $CD8^+$ T cells express high levels of activation markers, contain multiple phosphorylated proteins consistent with chronic activation, and secrete high levels of effector cytokines. The autoimmune syndrome in these animals is T cell-dependent, as interbreeding of CTLA-4-deficient mice with $TCRa^{-/-}$ or $RAG2^{-/-}$ mice prevents the disease. In addition, introduction of a TCR transgene onto this background does not restore immunity, suggesting that the T cells mediating the lymphoproliferation must be specific for particular antigens.[44-46] Two other manipulations have also been successful at preventing lymphoproliferation in CTLA-4-deficient mice. Depletion of $CD4^+$ but not $CD8^+$ T cells from birth prevented the onset of disease, indicating the primary importance of $CD4^+$ T cells in initiating this autoimmune process.[47] Similarly, interference with CD28-dependent costimulation prevented lymphoproliferation in $CTLA4^{-/-}$ mice. This was accomplished either by treatment of the mice with CTLA-4-Ig (a soluble form of CTLA-4 that binds B7 family members with high affinity and prevents CD80 and CD86 from ligating CD28) or by interbreeding CTLA-4-deficient mice with CD80/CD86 double-deficient animals.[48,49] Taken together, these experiments indicate that specific TCR and CD28 signals are essential to trigger the initial activation of $CD4^+$ and, secondarily, of $CD8^+$ T cells, and that the normal function of CTLA-4 is to prevent this proliferation at some level.

Experiments using TCR transgenic/$RAG2^{-/-}$/$CTLA4^{-/-}$ T cells have shown that reduced numbers of stimulator cells are necessary to promote proliferation and cytokine production by CTLA-4-deficient compared to CTLA-4-expressing T cells. In addition, the greater cytokine production observed by $CTLA\text{-}4^{-/-}$ T cells resulted from a greater number of cells being activated rather than from an increased magnitude of cytokine production per cell (Gajewski et al manuscript submitted). Taken together, these data indicate that CTLA-4 ligation raises the threshold for TCR-mediated T cell activation. These observations fit nicely with those made with CD28, ligation of which appears to lower TCR signaling threshold.[50]

Effects of CTLA-4 in Different T Cell Subsets and Stages of T Cell Differentiation

Both $CD4^+$ and $CD8^+$ T cells express high levels of CTLA-4 following activation. However, in $CTLA\text{-}4^{-/-}$ mice, activation markers are predominantly expressed on $CD4^+$ T cells. Moreover, depletion of $CD4^+$ T cells from birth prevents the autoimmune disease observed in these animals. Taken together, these results initially suggested that CTLA-4 may have a more critical inhibitory function for $CD4^+$ than for $CD8^+$ T cells. However, other experiments have shown that CTLA-4 clearly can inhibit $CD8^+$ T cells, as exemplified by studies performed using $CD8^+$ TCR-transgenic T cells in vitro, in the absence of $CD4^+$ cells.[44,46,51] Engagement of CTLA-4 by CD80 in these $CD8^+$ cells has been shown to result in reduced antigen-mediated proliferation and cytokine production. Additionally, blockade of CTLA-4 in other models in vivo has been shown to result in increased proliferation and cytolytic activity of $CD8^+$ T cells, in a CD4-independent manner.[52]

One hypothesis that can reconcile these discrepancies is that, in the naive state, $CD4^+$ but not $CD8^+$ T cells might be susceptible to regulation by CTLA-4, whereas in the primed effector state, both subsets might be affected. Indeed, evidence has indicated that primed T cells are in general more effectively inhibited by CTLA-4 compared to naive T cells.[45,46] However, some studies have shown that naive $CD4^+$ T cells can be inhibited by CTLA-4 ligation, whereas other studies have found that this was not the case for naive $CD8^+$ T cells. A hypothetical model suggesting the differential regulation of $CD4^+$ and $CD8^+$ T cells by CTLA-4 is diagramed in Figure 2. Confirmation of this model will require direct comparison of naive versus primed $CD4^+$ and $CD8^+$ T in the same experiments.

Following initial activation in response to specific antigen, naive T cells can proliferate and differentiate into one of at least two distinct functional phenotypes. T cells that secrete IFN-γ and IL-2 upon restimulation are termed Th1 cells, whereas T cells that secrete IL-4, IL-5, IL-6, and IL-10 are termed Th2 cells.[53] Although expression of CTLA-4 appeared to be markedly higher in Th2 compared to Th1 $CD4^+$ T cell clones, CTLA-4 cross-linking led to reduced production of cytokines by both subsets, including IFN-γ, TNF-α, IL-3, IL-4, IL-5, and IL-10.[54] In contrast to this inhibitory effect on most cytokines, other studies have indicated that cross-linking of CTLA-4 may actually promote the production of TGF-β by Th1 and Th2 clones at the same time as it inhibits IL-2 or IL-4 production, respectively .[55] In fact, blockade of TGF-β reduced the inhibitory effect of CTLA-4 cross-linking, suggesting that the effects of ligating CTLA-4 may partially be mediated via the secondary action of TGF-β.

It has been suggested that CD28 costimulation is especially critical for Th2 differentiation.[56] Consistent with this notion, levels of IL-4 production were particularly elevated in T cells from CTLA-$4^{-/-}$ mice, suggesting that lack of CTLA-4 may have allowed preferential differentiation into Th2 cells.[57] Similar results were obtained using T cells from TCR transgenic $CD4^+$/CTLA-4-deficient mice.[58] Consistent with these data, CTLA-4 ligation may preferentially skew towards Th1 differentiation of $CD4^+$ T cells, perhaps in part via the secretion of TGF-β.[59] However, as reviewed above, ligation of CTLA-4 affects cytokine production by both Th1 and Th2 cells. A balanced interpretation of these apparently discrepant results is that during T cell differentiation, CTLA-4 may preferentially favor the acquisition of a Th1 phenotype, whereas following differentiation, CTLA-4 may regulate both Th1 and Th2 effector cells. This model has yet to be tested formally.

Signaling Properties of CTLA-4

Three major hypotheses have been proposed to explain the inhibitory mechanism of CTLA-4 in T cells (Fig. 3). First, because of its high affinity for B7 family members, it has been suggested that surface CTLA-4 may scavenge CD80 and CD86 ligands, thereby preventing ligation of CD28 and blocking costimulation of T cells. Second, cross-linking of CTLA-4 may actively deliver inhibitory signals that counter positive signals delivered

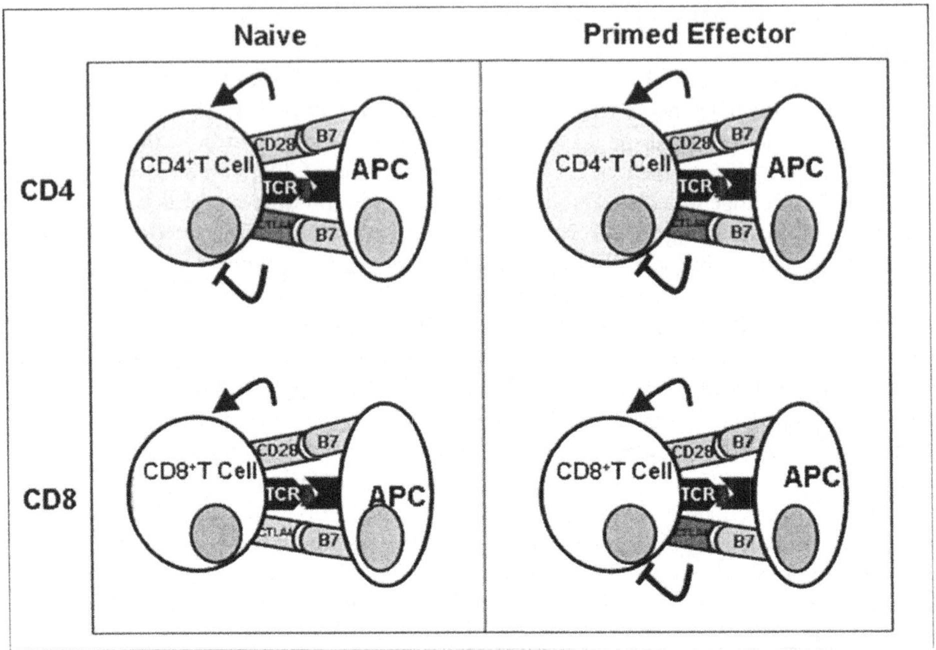

Figure 2. Model suggesting that CTLA-4 cross-linking inhibits naive and primed $CD4^+$ T cells, but only primed $CD8^+$ T cells.

by either CD28 or the TCR. Third, recent evidence has indicated that CTLA-4 is constitutively expressed by a subset of T cells that co-expresses CD25 and CD4. This subset contains so-called regulatory T cells capable of inhibiting the function of $CD25^-$ $CD4^+$ T cells. The presence of CTLA-4 on regulatory T cells has been shown to be crucial for their suppressor function.[43,60] This latter mechanism will be discussed in more detail below in the context of T cell tolerance.

Several pieces of evidence indicate that both scavenging of B7 molecules and signaling by CTLA4 do occur. Transgenic mice expressing a variant of CTLA-4 that lacks the cytoplasmic tail have been crossed onto a CTLA-4-deficient background.[61] T cells from these animals express CTLA-4 molecules that can bind to B7 and scavenge it from CD28 but presumably cannot signal. These mice exhibit an attenuated form of the autoimmune disease normally displayed by CTLA-4-deficient animals, indicating that scavenging B7 ligands provides a partial explanation for the inhibitory function of CTLA-4. In contrast, mice transgenic for a full-length version of CTLA-4 were completely protected from the autoimmune syndrome, attributing an important function to the cytoplasmic tail of CTLA-4. Similarly, in vitro experiments using cell-lines transfected with human CTLA-4 have shown that under conditions of B7-independent costimulation, inhibition of IL-2 production following CTLA-4 engagement required the CTLA-4 cytoplasmic domain. In contrast, under B7-dependent costimulation, inhibition of IL-2 production by CTLA-4 engagement was directly proportional to the levels of cell surface CTLA-4 and did not require its cytoplasmic region.[62]

The cytoplasmic tail of CTLA-4 contains two tyrosine residues but lacks intrinsic enzymatic activity, indicating that any signals mediated must occur indirectly via its

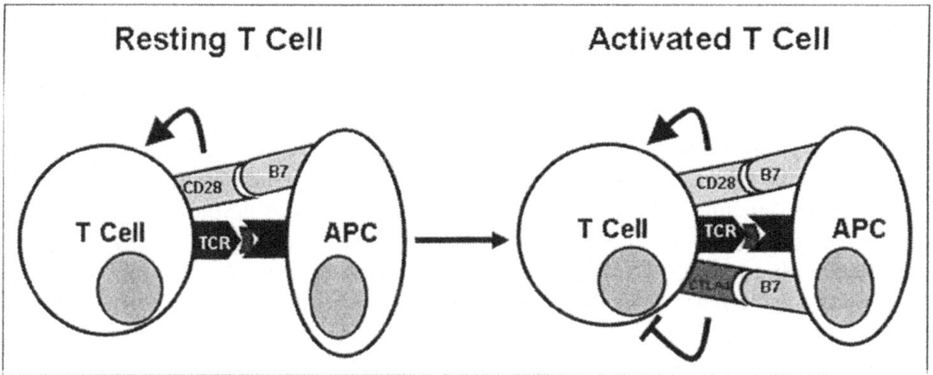

Figure 3. Mechanisms of T cell inhibition by CTLA-4. 1. CTLA-4 may scavenge B7 family members away from the costimulatory receptor CD28. 2. Ligation of CTLA-4 may deliver inhibitory signals to the T cell. 3. Inhibition of T cell function by CTLA-4 may depend on CTLA-4 conferring suppressor capacity to regulatory $CD4^+$ T cells. These mechanisms are not mutually exclusive.

association with intracellular enzymes. We have recently shown that TCR-mediated activation of normal T cells results in phosphorylation of endogenous CTLA-4 molecules (Gajewski et al, manuscript submitted), suggesting that tyrosine phosphorylation may be physiologically meaningful. Several tyrosine kinases have been shown capable of phosphorylating the cytoplasmic tyrosine residues of CTLA-4 in transfection experiments. These include the resting lymphocyte kinase Rlk, the Janus kinase JAK-2, and the Src family tyrosine kinases Fyn, Lyn, and Lck.[63-65] These events are specific, as the Syk-family kinase ZAP-70 was unable to promote CTLA-4 phosphorylation.

Because phosphorylation prevents CTLA-4 endocytosis, the subset of phosphorylated CTLA-4 might be expected to be retained at the cell surface, potentially to be ligated by B7 family members resulting in inhibition of T cell activation. Whether downstream signaling events depend upon this initial tyrosine phosphorylation of CTLA-4 remains a subject of investigation.

Several molecules have been found to associate with the cytoplasmic tail of CTLA-4 and have been proposed to mediate its downstream signaling. However, it is unclear whether any of them are necessary for the inhibitory effects mediated by CTLA-4 in normal T cells. The same YVKM motif that, in its unphosphorylated form, allows association with AP-50 and facilitates CTLA-4 endocytosis has been shown to promote interaction with several enzymes. The lipid kinase phosphatidylinositol 3-kinase (PI3K) is one such molecule.[66] Based on competition by synthetic peptides of the CTLA-4 tail, most investigators have found PI3K binding to CTLA-4 when the YVKM motif of CTLA-4 is phosphorylated on the tyrosine. However, one report has revealed association of endogenous PI3K with transfected full length CTLA-4 even in the absence of tyrosine phosphorylation.[65] Nevertheless, because inhibitors of PI3K activation do not appear to counter the inhibitory function of CTLA-4, it is unclear that this association is functionally relevant.

On the basis that a number of enzymes known to be important in TCR-mediated signaling are hyperphosphorylated in T cells from CTLA-4-deficient mice (including CD3z, ZAP70, SHC, Fyn, and Lck),[67] it has been proposed that CTLA-4 may recruit tyrosine phosphatases to the TCR. The tyrosine phosphatase SHP-1 has been found to bind phospho-polypeptides of the cytoplasmic tail of CTLA-4.[68] However, the inhibitory effects of CTLA-4 are intact in T cells from mice expressing reduced levels of SHP-1. The related tyrosine phosphatase SHP-2 has also been shown to associate with the PI3K-binding motif of CTLA-4,[67] although this interaction may be indirect as CTLA-4 lacks the typical sequence that can interact with SHP-2 SH2 domains.[69] However, recent evidence has shown that CTLA-4 can co-immunoprecipitate with CD3z and SHP-2 in a transfection/overexpression model, supporting the hypothesis that CTLA-4 may affect TCR signals directly by bringing SHP-2 to the TCR complex and facilitating dephosphorylation of proximal signaling molecules.[70] Although initial reports using phosphopeptides had suggested that the interaction of CTLA-4 with SHP-2 required phosphorylation of the tyrosine residue within the YVKM motif, more recent data have indicated that mutations of the 2 tyrosine residues in the cytoplasmic tail of CTLA-4 do not prevent CTLA-4 inhibitory effects[71-73] and do not abolish the interaction with SHP-2.[70] It is important to point out that most data available on CTLA-4 signaling have been obtained using transfected tumor cell-lines, and it is unclear whether results are going to hold true in normal T cells. For example, one report demonstrated that CTLA-4 cross-linking on normal preactivated T cells had no effect on phosphorylation of CD3z or ZAP-70.[74] Furthermore, although CTLA-4 cannot inhibit T cell function of naive $CD8^+$ T cells, endogenous SHP-2 is still found to co-immunoprecipitate with CTLA-4 in these cells, suggesting that association with SHP-2 may not be sufficient for the inhibitory effect of CTLA-4 (Gajewski et al, manuscript submitted).

Finally, results from a yeast-2 hybrid screening of a T cell library using the cytoplasmic tail of CTLA-4 as a bait have pointed to a new family of molecules, the PP2a family of serine/threonine phosphatases as potential candidates that may associate with CTLA-4.[75] Future studies may reveal whether these interactions play a role in CTLA-4 function.

Therefore, the mechanism by which the cytoplasmic tail of CTLA-4 exerts its inhibitory function is not fully understood. Nevertheless, CTLA-4 cross-linking has been show to inhibit activation of the MAP kinases JNK and ERK,[74] and to result in reduced activation of several transcription factors including NF-kB, NF-AT, and AP-1.[76,77]

Whether CTLA-4 antagonizes CD28 or TCR signals has been addressed by investigating the function of CTLA-4 in CD28-deficient T cells. Clear evidence has been generated demonstrating that CTLA-4 can inhibit T cell responses in the absence of CD28 both in vitro and in vivo.[51,78] In fact, blockade of CTLA-4 in vivo results in acceleration of cardiac allograft rejection in CD28-deficient mice. Finally, under conditions in which ligation of CTLA-4 in human T cells inhibits IL-2 production, no effect is observed on upregulation of $Bclx_L$, a member of the Bcl-2 family of anti-apoptotic genes that is induced by CD28-mediated signals.[79] Therefore, CTLA-4 engagement can inhibit signals delivered by the TCR and does not affect all signals triggered by CD28 ligation.

Inhibitory Effects of CTLA-4 in Experimental Models of Disease

Hints regarding the importance of CTLA-4 in immune responses in vivo, as well as its identification as a potential target for treatment of immunopathologic conditions, has come from studies of CTLA-4 blockade in experimental models of disease. Early studies showed that treatment of mice with anti-CTLA-4 mAb could promote rejection of transplantable tumors in vivo.[80] These experiments indicated that CTLA-4 blockade could enhance in vivo immune responses. In fact, this treatment was effective even when tumors were pre-established, a situation that more closely mimics the stage at which cancer patients would be treated. In other models, anti-CTLA-4 mAb has shown synergistic therapeutic effects when combined with immunization with GM-CSF-secreting tumor cells,[81] low doses of chemotherapeutic agents,[82] or with surgical resection of tumors[83] The mechanism by which anti-CTLA-4 mAbs promote tumor rejection may vary with each model system, as different tumors may utilize distinct mechanisms to evade immune responses. Interestingly, one report suggested that blockade of CTLA-4 reversed a tumor-induced non-responsive state of $CD8^+$ T cells in a CD4- and IL-2-dependent manner.[84]

In contrast to the beneficial outcome resulting from potentiation of anti-tumor immune responses with anti-CTLA-4 mAb, other systems have shown exacerbation of T cell-mediated autoimmune diseases and other settings worsened by increased T cell function. Thus, treatment with anti-CTLA-4 mAb has been shown to aggravate clinical disease in a model of experimental autoimmune encephalomyelitis,[85] as well as to worsen cutaneous leishmaniasis in susceptible mice probably due to a marked increase in IL-4 production.[86] In addition, anti-CTLA-4 mAb treatment was reported to increase the incidence and severity of diabetes in a TCR transgenic model in which T cells are specific for a pancreatic islet antigen.[87] This model has also allowed confirmation in vivo of the fact that the predominant effect of CTLA-4 occurs in primed T cells. Exacerbation of diabetes induced by adoptive transfer of diabetogenic TCR transgenic T cells was not observed when CTLA-4 was blocked during initial T cell activation in draining pancreatic lymph nodes, but rather when CTLA-4 was blocked during re-encounter with antigen after infiltration of the pancreas.[88]

Furthermore, CTLA-4 blockade has been reported to protect mice from infectious agents such as nematodes or cryptococcus,[89,90] suggesting that immune potentiation by anti-CTLA-4 mAb may have clinical utility in the treatment of certain infections.

Role of CTLA-4 in Central and Peripheral Tolerance

In addition to the process of negative selection by which autoreactive T cells are deleted in the thymus, several peripheral mechanisms appear to operate to prevent T cell reactivity against self-antigens. These include anergy, clonal deletion, immune deviation, and active suppression. Evidence has accumulated suggesting that CTLA-4 may play a role in regulating each of these mechanisms of T cell tolerance.

CTLA-4-deficient mice have normal thymocyte development indicating that CTLA-4 is not absolutely required for negative selection.[91] However, CTLA-4 is expressed in thymocytes upon TCR stimulation, and direct injection of anti-CTLA-4 mAb in the thymus has been reported to block thymocyte deletion induced by systemic injection of anti-CD3 mAb.[92] This result suggests that CTLA-4 may favor negative selection in the thymus. Furthermore, intrathymic injection of antigens can also result in peripheral tolerance to those antigens, both by deletion of thymocytes and induction of anergy. CTLA-4 blockade was recently shown to abrogate the induction tolerance induced by intrathymic injection of myelin antigens in a model of autoimmune encephalomyelitis.[93] Interestingly, anti-CTLA-4 treatment was not effective at breaking tolerance once established and had, therefore, a restricted window of action after priming with antigen. This indicates that CTLA-4 may be required for the induction but not the maintenance of thymic tolerance.

Several reports have indicated that CTLA-4 is necessary for the induction of peripheral tolerance in different models involving $CD4^+$ T cells. For instance, blockade of CTLA-4 in vivo has been shown to prevent tolerance induced by systemic injection of ovalbumin peptide in mice transfused with ovalbumin-specific $CD4^+$ TCR transgenic T cells, suggesting that CTLA-4 signals are necessary to achieve a tolerant state.[94] Similar results have been obtained in models of oral tolerance in which feeding of high doses of antigen results in peripheral T cells with reduced proliferative capacity to that antigen.[95] Tolerance in this case is thought to result both from depletion of antigen-specific T cells as well as anergy of residual cells. The exact pathway controlled by CTLA-4 remains to be investigated in this model. CTLA-4 was also necessary for anergy induction of ovalbumin-specific CD4$^+$ T cells in vitro, in a manner depending on ligation to CD80 (17). These results confirm the importance of CTLA-4 for induction of anergy in CD4$^+$ transgenic T cells, and further suggest that CD80 is the primary ligand for tolerance induction by CTLA-4. In a tumor model in CD4$^+$ TCR transgenic mice, however, blockade of CTLA-4 was shown to enhance the priming of responsive T cells but fail to prevent the induction of tumor antigen-specific tolerance.[96]

In contrast to most of the above results obtained with CD4$^+$ T cells, CTLA-4$^{-/-}$ CD8$^+$ TCR transgenic T cells have been shown to remain susceptible to anergy-induction by anti-CD3 mAb in vitro, suggesting that CTLA-4 is not required for this type of tolerance in this model[97] Whether the differences between these systems depend on the T cell subset, the stage of T cell differentiation, the specific TCR transgene expressed, or on the APCs present in the system, remain to be investigated.

Blockade of CTLA-4 can also prevent prolongation of skin graft survival induced by the combination of anti-CD80 and anti-CD86 mAbs, or by treatment with anti-CD154 mAb combined with donor splenocyte transfusion (98). In the latter case, CTLA-4 blockade has been shown to prevent deletion of alloreactive CD8$^+$ T cells.

Anti-CTLA-4 mAb has also been shown to increase the expansion and reduce the subsequent unresponsiveness of T cells following injection of staphylococcal enterotoxin B (SEB).[99] In addition, T cells isolated from these mice produced increased IL-4 upon restimulation in vitro, supporting a role for CTLA-4 in promoting immune deviation and type-2 T cell differentiation.

Finally, CTLA-4 may play a major role in the function of suppressor cells. The first evidence that a subset of cells could control CTLA-4-deficient T cells and prevent the disease normally observed in CTLA-4-deficient animals came from mixed bone marrow chimera studies.[100] Transfusion of bone marrow from CTLA-4-deficient mice into RAG1-deficient mice resulted in T cell infiltration of different organs, similar to that found in CTLA-4-deficient mice. However, if normal bone marrow was mixed with CTLA-4-deficient bone marrow, the disease did not occur, indicating a suppressive function of the wildtype bone marrow. Significant attention has been given the past few years to a subset of CD4$^+$ T cells that express high CD25 or low CD45RB and that appear to exert inhibitory activity for normal resting T cells. This subpopulation is also positive for CTLA-4 and, when sorted, can inhibit the proliferation and IL-2 production of CD25$^-$ CD4$^+$ T cells in a CTLA-4-dependent manner.[43,60] The mechanism of this inhibition is not fully understood, but may include secretion of TGF-β and modification of APCs that in turn induce tolerance to fresh T cells. This suppressor population appears to play a crucial role in prevention of autoimmune diseases. For example, CD28-deficient NOD mice that have markedly reduced numbers of CD25$^+$ CD4$^+$ T cells develop an accelerated form of diabetes, when compared with CD28-expressing NOD mice. Disease can be prevented by injection of CD25$^+$ CD4$^+$ T cells, supporting a regulatory activity within this subset of cells.[101] Similarly, adoptive transfer of CD25$^+$ T cells has been shown to prevent the onset of inflammatory bowel disease induced by transfer of CD45RB^{high} CD4$^+$ T cells to RAG1-deficient mice.[60] This protection was prevented by CTLA-4 blockade in vivo, indicating a major function of CTLA-4 in the inhibitory function of the $CD25^{+}$ $CD4^{+}$ T cell subset.

Role of CTLA-4 in Autoimmune Diseases in Humans

The importance of CTLA-4 in regulating autoimmune diseases in humans has come from the observation that expression of specific CTLA-4 polymorphisms correlates with a higher incidence of certain autoimmune diseases.[102-105] Polymorphisms have been identified in 3 regions of the CTLA-4 gene. These include the promoter region of the gene at position 318 from the ATG start codon, another at position 49 in the first exon encoding for CTLA-4, and a third in an (AT)n repeat within the 3'-untranslated region of exon 3. Expression of distinct forms of each of these has been associated with specific autoimmune disorders. However, the mechanism by which CTLA-4 polymorphisms contribute to the pathogenesis of each disease remains unclear. Recently, the first evidence that a specific polymorphism correlates with reduced inhibitory function of CTLA-4 in human T cells has been reported.[106] Patients with autoimmune Graves' disease had a higher frequency of G/G alleles and lower frequency of A/A alleles at position 49 compared to control patients. The presence of G/G alleles correlated with increased T cell proliferation following alloantigenic stimulation in vitro when compared with that of T cells expressing A/A alleles. In addition, CTLA-4 exerted less profound inhibitory effects on T cell proliferation in subjects bearing the G/G rather than the A/A genotype. These results indicate a correlation between the CTLA-4 genotype and the function of CTLA-4 protein in T cells, and suggest a possible mechanism by which these T cells may display autoreactivity in vivo.

Conclusion

In the last several years, tremendous progress has been made in the understanding of the functional properties of CTLA-4 and its potential role in regulating the immune response as it impacts on the onset or progression of certain diseases. Further work will be necessary to unveil fully the biochemical mechanisms by which CTLA-4 exerts its inhibitory activity. However, sufficient information has been gathered to suggest that CTLA-4 could be an important therapeutic target for the treatment of immunologically relevant diseases in man. Anti-CTLA-4 mAbs may prove useful in the induction of anti-tumor immunity or of improved defense against certain pathogens. In contrast, strategies to promote CTLA-4 ligation could be used to suppress unwanted autoimmunity or immune responses against transplanted allografts. The first clinical trials of anti-human CTLA-4 mAbs in human cancer patients are soon to be initiated. Alternative approaches to augment CTLA-4 ligation are now being explored in pre-clinical models.

References

1. Freeman GJ, Long AJ, Iwai Y et al. Engagement of the PD-1 immunoinhibitory receptor by a novel B7 family member leads to negative regulation of lymphocyte activation [In Process Citation]. J Exp Med 2000; 192:1027-1034.
2. Brunet JF, Denizot F, Luciani MF et al. A new member of the immunoglobulin superfamily—CTLA-4. Nature 1987; 328:267-270.
3. Howard TA, Rochelle JM, Seldin MF. Cd28 and Ctla-4, two related members of the Ig supergene family, are tightly linked on proximal mouse chromosome 1. Immunogenetics 1991; 33:74-76.
4. Lafage-Pochitaloff M, Costello R, Couez D et al. Human CD28 and CTLA-4 Ig superfamily genes are located on chromosome 2 at bands q33-q34. Immunogenetics 1990; 31:198-201.
5. Lenschow DJ, Walunas TLBluestone JA. CD28/B7 system of T cell costimulation. Ann Rev Immunol 1996; 14:233-258.

6. Borriello F, Sethna MP, Boyd SD et al. B7-1 and B7-2 have overlapping, critical roles in immunoglobulin class switching and germinal center formation. Immunity 1997; 6:303-313.
7. Linsley PS, Nadler SG, Bajorath J et al. Binding stoichiometry of the cytotoxic T lymphocyte-associated molecule-4 (CTLA-4). A disulfide-linked homodimer binds two CD86 molecules. J Biol Chem 1995; 270:15417-15424.
8. Greene JL, Leytze GM, Emswiler J et al. Covalent dimerization of CD28/CTLA-4 and oligomerization of CD80/CD86 regulate T cell costimulatory interactions. J Biol Chem 1996; 271:26762-26771.
9. Ostrov DA, Shi W, Schwartz JC et al. Structure of murine CTLA-4 and its role in modulating T cell responsiveness. Science 2000; 290:816-819.
10. Peach RJ, Bajorath J, Brady W et al. Complementarity determining region 1 (CDR1)- and CDR3-analogous regions in CTLA-4 and CD28 determine the binding to B7-1. J Exp Med 1994; 180:2049-2058.
11. Ellis JH, Burden MN, Vinogradov DV et al. Interactions of CD80 and CD86 with CD28 and CTLA4. J Immunol 1996; 56:2700-2709.
12. Morton PA, Fu XT, Stewart JA et al. Differential effects of CTLA-4 substitutions on the binding of human CD80 (B7-1) and CD86 (B7-2). J Immunol 1996; 156:1047-1054.
13. Nabavi N, Freeman GJ, Gault A et al. Signalling through the MHC class II cytoplasmic domain is required for antigen presentation and induces B7 expression. Nature 1992; 360:266-268.
14. Stack RM, Lenschow DJ, Gray GS et al. IL-4 treatment of small splenic B cells induces costimulatory molecules B7-1 and B7-2. J Immunol 1994; 152:5723-5733.
15. Hollsberg P, Scholz C, Anderson DE et al. Expression of a hypoglycosylated form of CD86 (B7-2) on human T cells with altered binding properties to CD28 and CTLA-4. J Immunol 1997; 159:4799-4805.
16. Judge TA, Wu Z, Zheng XG et al. The role of CD80, CD86, and CTLA4 in alloimmune responses and the induction of long-term allograft survival. J Immunol 1999; 162:1947-1951.
17. Chai JG, Vendetti S, Amofah E et al. CD152 ligation by CD80 on T cells is required for the induction of unresponsiveness by costimulation-deficient antigen presentation. J Immunol 2000; 165:3037-3042.
18. Kuiper HM, Brouwer M, Linsley P et al. Activated T cells can induce high levels of CTLA-4 expression on B cells. J Immunol 1995; 155:1776-1783.
19. Pioli C, Gatta L, Ubaldi V et al. Inhibition of IgG1 and IgE production by stimulation of the B cell CTLA-4 receptor [In Process Citation]. J Immunol 2000; 165:5530-5536.
20. Magistrelli G, Jeannin P, Herbault N, Benoit De Coignac A, Gauchat JF, Bonnefoy JY, et al. A soluble form of CTLA-4 generated by alternative splicing is expressed by nonstimulated human T cells. Eur J Immunol 1999, 29:3596-3602.
21. Oaks MK, Hallett KM, Penwell RT et al. A native soluble form of CTLA-4. Cell Immunol 2000; 201:144-153.
22. Oaks MK, Hallett KM. Cutting edge: a soluble form of CTLA-4 in patients with autoimmune thyroid disease. J Immunol 2000; 164:5015-5018.
23. Walunas TL, Lenschow DJ, Bakker CY et al. CTLA-4 can function as a negative regulator of T cell activation. Immunity 1994; 1:405-413.
24. Alegre ML, Noel PJ, Eisfelder BJ et al. Regulation of surface and intracellular expression of CTLA4 on mouse T cells. J Immunol 1996; 157:4762-4770.
25. Lindsten T, Lee KP, Harris ES et al. Characterization of CTLA-4 structure and expression on human T cells. J Immunol 1993; 151:3489-3499.
26. Finn PW, He H, Wang Y et al. Synergistic induction of CTLA-4 expression by costimulation with TCR plus CD28 signals mediated by increased transcription and messenger ribonucleic acid stability. J Immunol 1997; 158:4074-4081.
27. Chuang E, Alegre ML, Duckett CS et al. Interaction of CTLA-4 with the clathrin-associated protein AP50 results in ligand-independent endocytosis that limits cell surface expression. J Immunol 1997; 159:144-151.
28. Shiratori T, Miyatake S, Ohno H et al. Tyrosine phosphorylation controls internalization of CTLA-4 by regulating its interaction with clathrin-associated adaptor complex AP-2. Immunity 1997; 6:583-589.
29. Bradshaw JD, Lu P, Leytze G et al. Interaction of the cytoplasmic tail of CTLA-4 (CD152) with a clathrin-associated protein is negatively regulated by tyrosine phosphorylation. Biochemistry 1997; 36:15975-15982.

30. Chuang E, Alegre ML, Duckett CS et al. Interaction of CTLA-4 with the clathrin-associated protein AP50 results in ligand-independent endocytosis that limits cell surface expression. J Immunol 1997; 159:144-151.
31. Schneider H, Martin M, Agarraberes FA et al. Cytolytic T lymphocyte-associated antigen-4 and the TCR zeta/CD3 complex, but not CD28, interact with clathrin adaptor complexes AP-1 and AP-2. J Immunol 1999; 163:1868-1879.
32. Iida T, Ohno H, Nakaseko C et al. Regulation of cell surface expression of CTLA-4 by secretion of CTLA-4-containing lysosomes upon activation of CD4(+) T cells. J Immunol 2000; 165:5062-5068.
33. Krummel MF, Allison JP. CD28 and CTLA-4 have opposing effects on the response of T cells to stimulation. J Exp Med 1995; 182:459-465.
34. Walunas TL, Bakker CY, Bluestone JA. CTLA-4 ligation blocks CD28-dependent T cell activation. J Exp Med 1996; 183:2541-2550.
35. Krummel MF, Allison JP. CTLA-4 engagement inhibits IL-2 accumulation and cell cycle progression upon activation of resting T cells. J Exp Med 1996; 183:2533-2540.
36. Gribben JG, Freeman GJ, Boussiotis VA et al. CTLA4 mediates antigen-specific apoptosis of human T cells. Proc Natl Acad Sci USA 1995; 92:811-815.
37. Scheipers PReiser H. Fas-independent death of activated CD4+ T lymphocytes induced by CTLA-4 crosslinking. Proc Natl Acad Sci USA 1998; 95:10083-10088.
38. Brunner MC, Chambers CA, Chan FK et al. CTLA-4-Mediated inhibition of early events of T cell proliferation. J Immunol 1999; 162:5813-5820.
39. Griffin MD, Hong DK, Holman PO et al. Blockade of T cell activation using a surface-linked single-chain antibody to CTLA-4 (CD152). J Immunol 2000; 164:4433-4442.
40. Linsley PS, Bradshaw J, Greene J et al. Intracellular trafficking of CTLA-4 and focal localization towards sites of TCR engagement. Immunity 1996; 4:535-543.
41. Waterhouse P, Penninger JM, Timms E et al. Lymphoproliferative disorders with early lethality in mice deficient in Ctla-4. Science 1995; 270:985-988.
42. Tivol EA, Borriello F, Schweitzer AN et al. Loss of CTLA-4 leads to massive lymphoproliferation and fatal multiorgan tissue destruction, revealing a critical negative regulatory role of CTLA-4. Immunity 1995; 3:541-547.
43. Takahashi T, Tagami T, Yamazaki S et al. Immunologic self-tolerance maintained by CD25(+)CD4(+) regulatory T cells constitutively expressing cytotoxic T lymphocyte-associated antigen 4. J Exp Med 2000; 192:303-310.
44. Waterhouse P, Bachmann MF, Penninger JM et al. Normal thymic selection, normal viability and decreased lymphoproliferation in T cell receptor-transgenic CTLA-4-deficient mice. Eur J Immunol 1997; 27:1887-1892.
45. Chambers CA, Kuhns MSAllison JP. Cytotoxic T lymphocyte antigen-4 (CTLA-4) regulates primary and secondary peptide-specific CD4(+) T cell responses. Proc Natl Acad Sci USA 1999; 96:8603-8608.
46. Chambers CA, Sullivan TJ, Truong T et al. Secondary but not primary T cell responses are enhanced in CTLA-4-deficient CD8+ T cells. Eur J Immunol 1998; 28:3137-3143.
47. Chambers CA, Sullivan TJ, Allison JP. Lymphoproliferation in CTLA-4-deficient mice is mediated by costimulation-dependent activation of $CD4^+$ T cells. Immunity 1997; 7:885-895.
48. Tivol EA, Boyd SD, McKeon S et al. CTLA4Ig prevents lymphoproliferation and fatal multiorgan tissue destruction in CTLA-4-deficient mice. J Immunol 1997; 158:5091-5094.
49. Mandelbrot DA, McAdam AJ, Sharpe AH. B7-1 or B7-2 is required to produce the lymphoproliferative phenotype in mice lacking cytotoxic T lymphocyte-associated antigen 4 (CTLA-4). J Exp Med 1999; 189:435-440.
50. Viola ALanzavecchia A. T cell activation determined by T cell receptor number and tunable thresholds. Science 1996; 273:104-106.
51. Fallarino F, Fields PE, Gajewski TF. B7-1 engagement of cytotoxic T lymphocyte antigen 4 inhibits T cell activation in the absence of CD28. J Immunol 1998; 188:205-210.
52. McCoy KD, Hermans IF, Fraser JH et al. Cytotoxic T lymphocyte-associated antigen 4 (CTLA-4) can regulate dendritic cell-induced activation and cytotoxicity of CD8(+) T cells independently of CD4(+) T cell help. J Exp Med 1999; 189:1157-1162.
53. Singh VK, Mehrotra S, Agarwal SS. The paradigm of Th1 and Th2 cytokines: its relevance to autoimmunity and allergy. Immunol Res 1999; 20:147-161.

54. Alegre ML, Shiels H, Thompson CB et al. Expression and function of CTLA-4 in Th1 and Th2 cells. J Immunol 1998; 161:3347-3356.
55. Chen W, Jin W, Wahl SM. Engagement of cytotoxic T lymphocyte-associated antigen 4 (CTLA-4) induces transforming growth factor beta (TGF-beta) production by murine CD4(+) T cells. J Exp Med 1998; 188:1849-1857.
56. Rulifson IC, Sperling AI, Fields PE et al. CD28 costimulation promotes the production of Th2 cytokines. J Immunol 1997; 158:658-665.
57. Khattri R, Auger JA, Griffin MD et al. Lymphoproliferative disorder in CTLA-4 knockout mice is characterized by CD28-regulated activation of Th2 responses. J Immunol 1999; 162:5784-5791.
58. Oosterwegel MA, Mandelbrot DA, Boyd SD et al. The role of CTLA-4 in regulating Th2 differentiation. J Immunol 1999; 163:2634-2639.
59. Kato T, Nariuchi H. Polarization of Naive CD4+ T Cells Toward the Th1 Subset by CTLA-4 Costimulation. J Immunol 2000; 164:3554-3562.
60. Read S, Malmstrom V, Powrie F. Cytotoxic T Lymphocyte-associated Antigen 4 Plays an Essential Role in the Function of CD25(+)CD4(+) Regulatory Cells that Control Intestinal Inflammation. J Exp Med 2000; 192:295-302.
61. Masteller EL, Chuang E, Mullen AC et al. Structural analysis of CTLA-4 function in vivo. J Immunol 2000; 164:5319-5327.
62. Carreno BM, Bennett F, Chau TA et al. CTLA-4 (CD152) can inhibit T cell activation by two different mechanisms depending on its level of cell surface expression. J Immunol 2000; 165:1352-1356.
63. Miyatake S, Nakaseko C, Umemori H et al. Src family tyrosine kinases associate with and phosphorylate CTLA-4 (CD152). Biochem Biophys Res Commun 1998; 248:444-448.
64. Chikuma S, Murakami M, Tanaka K et al. Janus kinase 2 is associated with a box 1-like motif and phosphorylates a critical tyrosine residue in the cytoplasmic region of cytotoxic T lymphocyte associated molecule-4 [In Process Citation]. J Cell Biochem 2000; 78:241-250.
65. Chuang E, Lee KM, Robbins MD et al. Regulation of cytotoxic T lymphocyte-associated molecule-4 by Src kinases. J Immunol 1999; 162:1270-1277.
66. Schneider H, Prasad KVS, Shoelson SE et al. CTLA-4 binding to the lipid kinase phosphatidylinositol 3-kinase in T cells. J Exp Med 1995; 181:351-355.
67. Marengere LE, Waterhouse P, Duncan GS et al. Regulation of T cell receptor signaling by tyrosine phosphatase SYP association with CTLA-4. Science 1996; 272:1170-1173.
68. Chambers CA, Allison JP. The role of tyrosine phosphorylation and PTP-1C in CTLA-4 signal transduction. Eur J Immunol 1996; 26:3224-3229.
69. Schneider H, Rudd CE. Tyrosine phosphatase SHP-2 binding to CTLA-4: absence of direct YVKM/YFIP motif recognition. Biochem Biophys Res Commun 2000; 269:279-283.
70. Lee K-M, Chuang E, Griffin M et al.: Molecular basis of T cell inactivation by CTLA-4. Science 1998; 282:2262-2266.
71. Nakaseko C, Miyatake S, Iida T et al. Cytotoxic T lymphocyte antigen 4 (CTLA-4) engagement delivers an inhibitory signal through the membrane-proximal region in the absence of the tyrosine motif in the cytoplasmic tail. J Exp Med 1999; 190:765-774.
72. Cinek T, Sadra AImboden JB. Cutting edge: tyrosine-independent transmission of inhibitory signals by CTLA-4. J Immunol 2000; 164:5-8.
73. Baroja ML, Luxenberg D, Chau T et al. The inhibitory function of CTLA-4 does not require its tyrosine phosphorylation. J Immunol 2000; 164:49-55.
74. Revilla Calvo C, Amsen D, Kruisbeek AM. Cytotoxic T lymphocyte antigen 4 (CTLA-4) interferes with extracellular signal-regulated kinase (ERK) and jun NH2-terminal kinase (JNK) activation, but does not affect phosphorylation of T cell receptor zeta and ZAP70. J Exp Med 1997; 186:1645-1653.
75. Chuang E, Fisher TS, Morgan RW et al. The CD28 and CTLA-4 receptors associate with the serine/threonine phosphatase PP2A. Immunity 2000; 13:313-322.
76. Olsson C, Riebeck K, Dohlsen M et al. CTLA-4 ligation suppresses CD28-induced NF-kappaB and AP-1 activity in mouse T cell blasts. J Biol Chem 1999; 14:14400-14405.
77. Fraser JH, Rincon M, McCoy KD et al. CTLA4 ligation attenuates AP-1, NFAT and NF-kappaB activity in activated T cells. Eur J Immunol 1999; 29:838-844.
78. Lin H, Rathmell JC, Gray GS et al. Cytotoxic T lymphocyte antigen 4 (CTLA4) blockade accelerates the acute rejection of cardiac allografts in CD28-deficient mice: CTLA4 can function independently of CD28. J Exp Med 1998; 188:199-204.

79. Blair PJ, Riley JL, Levine BL et al. CTLA-4 ligation delivers a unique signal to resting human CD4 T cells that inhibits interleukin-2 secretion but allows Bcl-xL induction. J Immunol 1998; 160:12-15.
80. Leach DR, Krummel MF, Allison JP. Enhancement of antitumor immunity by CTLA-4 blockade. Science 1996; 271:1734-1736.
81. van Elsas A, Hurwitz AA, Allison JP. Combination immunotherapy of B16 melanoma using anti-cytotoxic T lymphocyte-associated antigen 4 (CTLA-4) and granulocyte/macrophage colony-stimulating factor (GM-CSF)-producing vaccines induces rejection of subcutaneous and metastatic tumors accompanied by autoimmune depigmentation. J Exp Med 1999; 190:355-366.
82. Mokyr MB, Kalinichenko T, Gorelik L et al. Realization of the therapeutic potential of CTLA-4 blockade in low-dose chemotherapy-treated tumor-bearing mice. Cancer Res 1998; 58:5301-5304.
83. Kwon ED, Foster BA, Hurwitz AA et al.: Elimination of residual metastatic prostate cancer after surgery and adjunctive cytotoxic T lymphocyte-associated antigen 4 (CTLA-4) blockade immunotherapy. Proc Natl Acad Sci USA 1999; 96:15074-15079.
84. Shrikant P, Khoruts A, Mescher MF. CTLA-4 blockade reverses CD8+ T cell tolerance to tumor by a $CD4^+$ T cell- and IL-2-dependent mechanism. Immunity 1999; 11:483-493.
85. Perrin PJ, Maldonado JH, Davis TA et al. CTLA-4 blockade enhances clinical disease and cytokine production during experimental allergic encephalomyelitis. J Immunol 1996; 157:1333-1336.
86. Heinzel FP, Maier RA Jr. Interleukin-4-independent acceleration of cutaneous leishmaniasis in susceptible BALB/c mice following treatment with anti-CTLA4 antibody. Infect Immun 1999; 67:6454-6460.
87. Luhder F, Hoglund P, Allison JP et al. Cytotoxic T lymphocyte-associated antigen 4 (CTLA-4) regulates the unfolding of autoimmune diabetes. J Exp Med 1998; 187:427-432.
88. Luhder F, Chambers C, Allison JP et al. Pinpointing when T cell costimulatory receptor CTLA-4 must be engaged to dampen diabetogenic T cells [In Process Citation]. Proc Natl Acad Sci USA 2000; 97:12204-12209.
89. McCoy K, Camberis M, Gros GL. Protective immunity to nematode infection is induced by CTLA-4 blockade. J Exp Med 1997; 186:183-187.
90. McGaha T, Murphy JW. CTLA-4 down-regulates the protective anticryptococcal cell-mediated immune response. Infect Immun 2000; 68:4624-4630.
91. Chambers CA, Cado D, Truong T et al. Thymocyte development is normal in CTLA-4-deficient mice. Proc Natl Acad Sci USA 1997; 94:9296-9301.
92. Cilio CM, Daws MR, Malashicheva A et al. Cytotoxic T lymphocyte antigen 4 is induced in the thymus upon in vivo activation and its blockade prevents anti-CD3-mediated depletion of thymocytes. J Exp Med 1998; 188:1239-1246.
93. Issazadeh S, Zhang M, Sayegh MH et al. Acquired thymic tolerance: role of CTLA4 in the initiation and maintenance of tolerance in a clinically relevant autoimmune disease model. J Immunol 1999; 162:761-765.
94. Perez VL, Van Parijs L, Biuckians A et al. Induction of peripheral T cell tolerance in vivo requires CTLA-4 engagement. Immunity 1997; 6:411-417.
95. Samoilova EB, Horton JL, Zhang H et al. CTLA-4 is required for the induction of high dose oral tolerance. Int.Immunol. 1998; 10:491-498.
96. Sotomayor EM, Borrello I, Tubb E et al. In vivo blockade of CTLA-4 enhances the priming of responsive T cells but fails to prevent the induction of tumor antigen-specific tolerance. Proc Natl Acad Sci USA 1999; 96:11476-11481.
97. Frauwirth KA, Alegre ML, Thompson CB. Induction of T cell anergy in the absence of CTLA-4/B7 interaction. J Immunol 2000; 164:2987-2993.
98. Markees TG, Phillips NE, Gordon EJ et al. Long-term survival of skin allografts induced by donor splenocytes and anti-CD154 antibody in thymectomized mice requires CD4(+) T cells, interferon-gamma, and CTLA4. J Clin Invest 1998; 101:2446-2455.
99. Walunas TL, Bluestone JA. CTLA-4 regulates tolerance induction and T cell differentiation in vivo. J Immunol 1998; 160:3855-3860.
100. Bachmann MF, Kohler G, Ecabert B et al. Cutting edge: lymphoproliferative disease in the absence of CTLA-4 is not T cell autonomous. J Immunol 1999; 163:1128-1131.
101. Salomon B, Lenschow DJ, Rhee L et al. B7/CD28 costimulation is essential for the homeostasis of the CD4+CD25+ immunoregulatory T cells that control autoimmune diabetes. Immunity 2000; 12:431-440.

102. Ligers A, Xu C, Saarinen S et al. The CTLA-4 gene is associated with multiple sclerosis. J Neuroimmunol 1999; 97:182-190.
103. Yanagawa T, Hidaka Y, Guimaraes V et al. CTLA-4 gene polymorphism associated with Graves' disease in a caucasian population. J Clin Endocrin Metab 1995; 80:41-45.
104. Vaidya B, Imrie H, Perros P et al. The cytotoxic T lymphocyte antigen-4 is a major Graves' disease locus. Hum Mol Genet 1999; 8:1195-1199.
105. Van der Auwera BJ, Vandewalle CL, Schuit FC et al. CTLA-4 gene polymorphism confers susceptibility to insulin-dependent diabetes mellitus (IDDM) independently from age and from other genetic or immune disease markers. The Belgian Diabetes Registry. Clin Exp Immunol 1997; 110:98-103.
106. Kouki T, Sawai Y, Gardine CA et al. CTLA-4 Gene Polymorphism at Position 49 in Exon 1 Reduces the Inhibitory Function of CTLA-4 and Contributes to the Pathogenesis of Graves' disease. J Immunol 2000; 165:6606-6611.

Chapter 3

CTLA-4 in Systemic Lupus Erythematosus

David I. Daikh and David Wofsy

Introducton

The recent characterization of several costimulatory interactions between antigen presenting cells and T cells represents a major advance in our understanding of both normal adaptive immune responses and pathologic autoimmune responses. Furthermore, characterization of these costimulation pathways has suggested a number of new approaches for directing more specific immunosuppressive therapy for chronic autoimmune diseases such as systemic lupus erythematosus (SLE). Many experiments have now established the validity the two-signal hypothesis of Bretscher and Cohn.[1] These studies have established that the first signal, provided by antigen specific stimulation of the T cell antigen receptor, is a necessary, but insufficient stimulus for maximal T cell activation and subsequent effector function. T cell activation, as manifested by IL-2 production and T cell proliferation, further requires a second, or costimulatory signal.[2,3] A variety of specific interactions between T cells and antigen presenting cells (APC), or other accessory cells, can provide this second signal. However, the interaction between CD28 expressed on T cells and B7 molecules expressed on antigen presenting cells and activated B cells appears to provide the major costimulatory signal to T cells in a wide variety of adaptive immune responses.[4] CD28 is constitutively expressed on the majority of T cells,[5] while its ligands are expressed on the surface of activated, but not resting APC.[6] Three B7 ligands have been described on activated human B lymphocytes; B7-1 (CD80), B7-2 (CD86) and B7-3, and two ligands, mB7-1 and mB7-2 have been identified of mouse B cells. Studies of CD28-deficient mice suggest that CD28 is the predominant receptor responsible for B7-dependent T cell activation.[7] More recently, an inducible costimulatory molecule called ICOS that is structurally related to CD28, along with its ligand B7h, that is also homologous to other B7 molecules, have been described.[8-10] These costimulatory molecules may deliver both complementary and unique costimulatory signals to T cells.[11,12]

CTLA4 is a T cell surface receptor that also binds B7 molecules and that has 20% sequence homology and significant structural homology to CD28.[13] However, CTLA4 binds both CD80 and CD86 with much higher avidity than does CD28 and in contrast to CD28, CTLA4 is only expressed on activated T cells. Although originally classified as a costimulatory molecule, it has become clear that a major function of CTLA4 is to negatively regulate T cell function.[14,15] The kinetics of CTLA4 induction and the high avidity of binding to B7 molecules imply that one role of CTLA4 is to limit T cell activation

CTLA-4 in Autoimmune Disease, edited by Flemming Pociot. ©2004 Eurekah.com.

that results from CD28-mediated costimulation. CTLA4 can also mediate antigen-specific apoptosis of T cells,[16] and antitumor immunity is enhanced following CTLA4 blockade.[17] Finally, in the complete absence of CTLA4, mice develop massive lymphoproliferation.[18] CTLA4 therefore appears to be critical for the regulation of lymphocyte function and for the limitation of immune responses, and can therefore be thought of as a coinhibitory molecule. However, the effects of CTLA4 engagement can differ depending upon the stage of an immune response at which it occurs. For example, a toleragenic stimulus can be converted into an abortive immune response by blocking CTLA4 with anti-CTLA4.[19] The importance of CTLA4 during the early phases of an immune response is emphasized by a study in a transgenic mouse model of autoimmune diabetes in which anti-CTLA4 can result in the rapid development of islet infiltration and diabetes many months before disease would normally develop. However, inflammation develops only when CTLA4 is administered during a narrow window of time (20). These experiments have emphasized that CTLA4 can also have an influence on the initial character of an immune response.

The high avidity of CTLA4 for B7 molecules suggested that a soluble form of CTLA4 would effectively block interactions between B7 molecules and CD28, thereby specifically blocking CD28-mediated costimulation. This strategy was utilized to make a recombinant CTLA4-Ig fusion protein, called CTLA4Ig. This molecule has been used in a variety of in vitro and in vivo systems to define the role of B7-CD28 interactions in normal and pathologic immune responses. In addition, studies using CTLA4Ig indicate that under certain experimental conditions, blockade of CD28 signaling can result in long term antigen-specific tolerance.[21,22] For example, CTLA4Ig was used to successfully prolong tissue allografts and xenografts in vivo.[23,24] Similarly, CTLA4Ig was used to induce long term unresponsiveness to induced autoantibody production in a Hepatitis B eAg transgenic mouse model.[25]

CLTA-4 in the Pathogenesis of Lupus

CLTA4Ig has also been very useful in studying the role of CD28-mediated costimulation in animal models of autoimmune disease, including SLE. Although there are a number of murine models of lupus, the (NZB x NZW) F_1 (B/W) mouse model most closely resembles SLE in humans with respect to disease manifestations and the development of more severe illness in females.[26] Like humans with SLE, B/W mice develop antibodies against native DNA (dsDNA) and immune-complex glomerulonephritis. Female mice begin to develop kidney disease, as manifest by proteinuria, beginning around five to six months of age. Autoantibodies against dsDNA and other nuclear components can be detected beginning at this time as well. Advanced kidney failure is the cause of death in most mice and mortality in this strain approaches 100% by one year of age. B/W mice also have impaired T cell function,[9-11] polyclonal B cell activation,[27-29] and defective immune clearance,[30] as is seen in human SLE. These similarities suggest that SLE in humans and in the B/W mouse share common pathogenic mechanisms. Additionally, a number of lines of evidence indicate that in both human and murine lupus, T cells are driving autoantibody production by B cells.[31] For example, $CD4^+$ T cells are necessary for the development dsDNA autoantibodies and disease in B/W mice.[32] The importance of T cell costimulation in SLE was suggested by studies in which CTLA4Ig administered to female B/W mice prevented the development of autoantibody production and kidney disease, resulting in prolonged survival.[33] Thus, in this murine model of SLE, T cell activation via CD28-B7 interactions is necessary for the progression of autoimmune disease and interruption of this pathway is protective. This conclusion has been strengthened by additional studies in which specific anti-B7 antibodies were administered to B/W mice. In this case, anti-B72 mAb inhibited IgG1 and IgG2b anti-dsDNA antibody production in B/W mice and reduced production of IL-2, IFNg, IL-4 and IL-6 by anti-CD3-stimulated splenocytes. However, inhibition of IgG2a dsDNA antibody production and the development of lupus nephritis required the presence of both anti-CD80 and anti-CD86 mAb.[34] A similar result was obtained using specific anti-B7 mAb in the MRL$^{lpr/lpr}$ lupus mouse model.[35] Anti-CD86 mAb treatment alone prevented anti-dsDNA, but not anti-small ribonucleoprotein antibodies. Both anti-CD80 and anti-CD86 were required to completely inhibit autoantibody production. Furthermore, MRL $^{lpr/lpr}$ mice genetically deficient in either CD80 or CD86 developed the same autoantibodies as wild-type mice. However, CD80-deficient mice developed more severe glomerulonephritis than CD86-deficient mice. These studies suggest that while each B7 costimulatory signal may have different effects on distinct immunopathological events in murine lupus, interaction of CD28 with either B7 molecule is sufficient for the activation of autoreactive T cells and the progression of autoimmune disease.

The importance the CD28-mediated T cell costimulation in murine lupus and the negative regulatory effect of CTLA4 on this pathway suggests the possibility that abnormalities of CTLA4 expression or function might contribute to excessive autoreactivity in SLE. Among patients with either active or inactive SLE there are generally fewer CD28$^+$ and increased CD28$^-$ circulating T cells.[36] In addition, some lupus patients may have a defect in the upregulation of B7 molecules on APC.[37] The significance of what would appear to be a propensity towards a decrease in T cell costimulation implied by these findings is not clear. However, since CD28$^+$ T cells from SLE patients had accelerated anti-CD3-induced apoptosis compared to controls, it was suggested that CD28-mediated costimulation, followed by CTLA4-induced apoptosis, might be involved in the T cell lymphopenia that is seen in SLE.[36] CTLA4 expression can be detected on peripheral T lymphocytes, but its expression level is low in both SLE patients and controls, although SLE patients have a greater percentage of CTLA4-positive cells than controls.[38] However, in vitro stimulation of SLE lymphocytes results in expression of CTLA4 after five days with the same kinetics as that observed in control cells.[36] A number of investigators have examined the CTLA4 gene locus to determine whether there are any associations between SLE and polymorphisms within this region. However, no such association has been demonstrated among cacausian, Mexican-American or Japanese cohorts.[39-42] Thus, to date, there is little evidence to suggest that abnormalities in coinhibition by CTLA4 underlie the pathogenesis of lupus.

Blockage of Costimulatory Signals in Murine Models of Human SLE

The opposing effects of CTLA4 and CD28 on T cell activation illustrate the general phenomenon of multiple signal integration in the regulation of the immune system. In the case of costimulatory interactions, not only are there self-limiting mechanisms, such as that provided by the upregulation of CTLA4 expression, but a number of different specific interactions between T cells and APC can serve to provide a costimulatory function. Furthermore, mutual regulatory interactions between these different costimulatory pathways occur. These complexities are illustrated by the costimulatory interaction between CD40L (CD40 ligand, CD154) and CD40. This interaction provides an important costimulatory signal to B cells, resulting in B cell proliferation, B cell maturation, increased Ig production and isotype switching.[43,44] Just as surface expression of CTLA4 is increased by T cell activation via CD28, CD40L expression on T cells is upregulated following T cell activation. CD40 engagement by CD40L in turn results in increased surface expression of B7 molecules on APC.[45]

Recent studies indicate that, as has been observed with agents such as CTLA4Ig that selectively block B7-CD28 interactions, agents that interrupt CD40-CD40L binding can suppress autoimmunity in murine models of lupus. For example, a short course of anti-CD40L can produce prolonged benefit in lupus-prone SWR/NZB F_1 (SNF_1) mice. Interestingly, treatment of SNF_1 mice with anti-CD40L suppressed lupus nephritis without eliminating pathogenic T cell clones, apparently by preventing autoantibody production by B cells.[46] Anti-CD40L was also effective in prolonging survival among SNF_1 mice with established nephritis, due to a decrease in renal inflammation.[47] To evaluate the relative importance of both the CD28-B7 and the CD40-CD40L costimulatory pathways in murine lupus, we have examined the effects of blocking both of these pathways simultaneously in B/W mice. When CTLA4Ig was combined with anti-CD40L, there was long lasting inhibition of autoantibody production and renal disease after only a two week course of treatment. This clinical benefit was significantly greater that that observed for either agent alone.[48] Taken together, these studies indicate that both of these costimulatory pathways are important in the stimulation of autoimmune reponses in murine lupus and that concomitant inhibition of these two pathways can produce synergistic benefit in blocking the progression of autoimmune pathology that lasts long after treatment has been discontinued.

CTLA4Ig Treatment in Murine Lupus Nephritis

The value of such an approach to treating autoimmunity is further supported by the observation that the therapeutic efficacy of interrupting T cell costimulation is comparable to that of conventional immunosuppressive agents. Cyclophosphamide is the most efficacious of the various available immunosuppressants and is standard therapy for active, diffuse proliferative lupus nephritis in humans.[49] In fact, its efficacy for lupus nephritis was first demonstrated in the B/W mouse model.[50,51] We have compared the efficacy of CTLA4Ig with that of cyclophosphamide in B/W mice and found that the effects on nephritis are similar.[52] We treated six month old B/W females with mild renal disease (proteinuria <100 mg/dl) with either CTX, CTLA4Ig, both agents concomitantly, or saline. 16 weeks after initiation of therapy, 80% of control mice had developed severe proteinuria, compared to 0% of mice treated with both CTX and CTLA4Ig (Fig. 1). Mice that had received either CTX or CTLA4Ig had progression of renal disease that was slower than that of controls. After treatment was stopped, there was progression of disease in all groups, such that by 20 weeks, the percentage of CTX-treated mice with severe proteinuria was similar to that in the CTLA4Ig group. However, the rate of progression of proteinuria among mice that received both CTLA4Ig and CTX was significantly lower. This delay in progression of proteinuria among mice that received either CTLA4Ig or CTX resulted in a prolongation in survival; 18 weeks after cessation of treatment, 27% of CTLA4Ig-treated mice and 36% of CTX-treated mice survived, compared with 8% of controls.

The beneficial effect of combining CTLA4Ig with CTX is even more apparent in advanced murine lupus nephritis. We found that although either CTX or CTLA4Ig can prolong survival of mice with advanced nephritis (proteinuria >300 mg/dl), the level of proteinuria is maintained at the pretreatment level. However, mice treated simultaneously

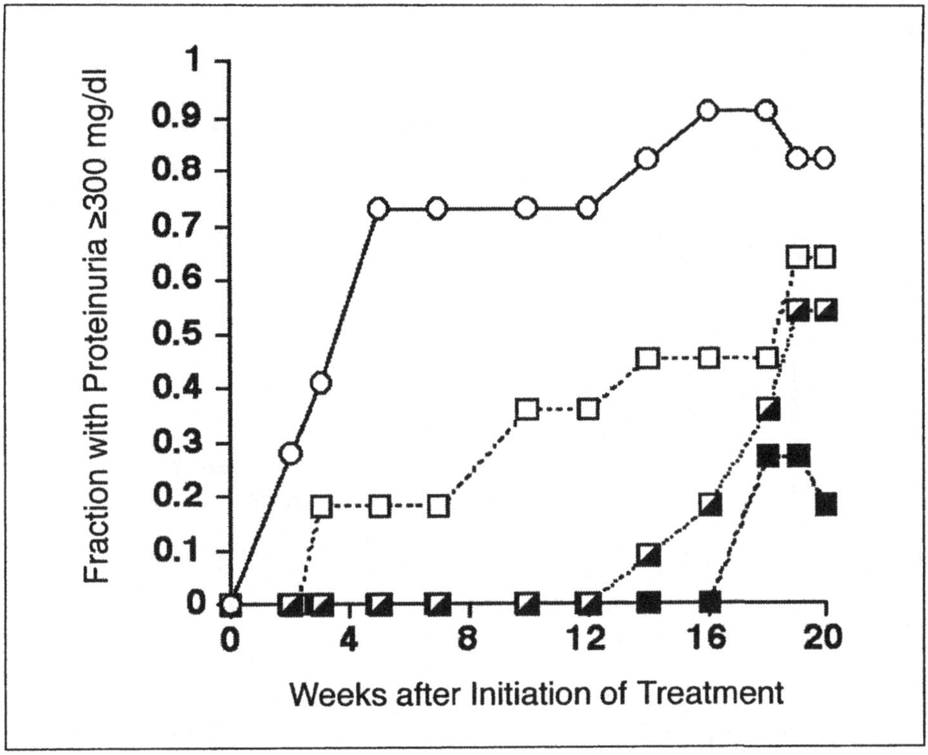

Figure 1. Development of proteinuria among mice with mild renal disease during and after 16 weeks of treatment with cyclophosphamide (half-solid boxes), CTLA4Ig (open boxes), combined cyclophosphamide and CTLA4Ig (solid boxes), or saline (open circles). Reprinted with permission from reference 46: Daikh, D.I. and D. Wofsy, Cutting edge: Reversal of murine lupus nephritis with CTLA4Ig and cyclophosphamide. J Immunology, 2001. 166: p. 2913-6. Copyright 2001. The American Association of Immunologists.

with both cyclophosphamide and CTLA4Ig exhibit a rapid and marked improvement in proteinuria that is sustained throughout the course of treatment, and that results in improved survival compared to mice treated with either agent alone (Fig. 2). Overall, we find that CTLA4Ig is comparable in efficacy to cyclophosphamide for the treatment of murine lupus, particularly lupus nephritis, in both early and advanced stages of disease. Thus, in the same model in which the effectiveness of CTX for lupus nephritis was first demonstrated, administration of an agent that blocks T cell costimulation has similar efficacy to CTX, the agent currently most widely used to treat human lupus nephritis. Furthermore, among animals with the most advanced disease, we find that the effects of these agents are complementary when used together.

These studies make it clear that T cell costimulation via CD28 plays a very important role in the activation of autoimmune responses in SLE and that interruption of this signal can have an ameliorative effect on autoimmune pathology. Regulatory mechanisms that impact the strength of this signal, such as that provided by CTLA4, are likely to also have a role in the overall strength of such autoimmune responses. To this extent, this also suggests that regulation of CTLA4 expression, or it's function, might provide a useful therapeutic target for autoimmune diseases such as lupus.

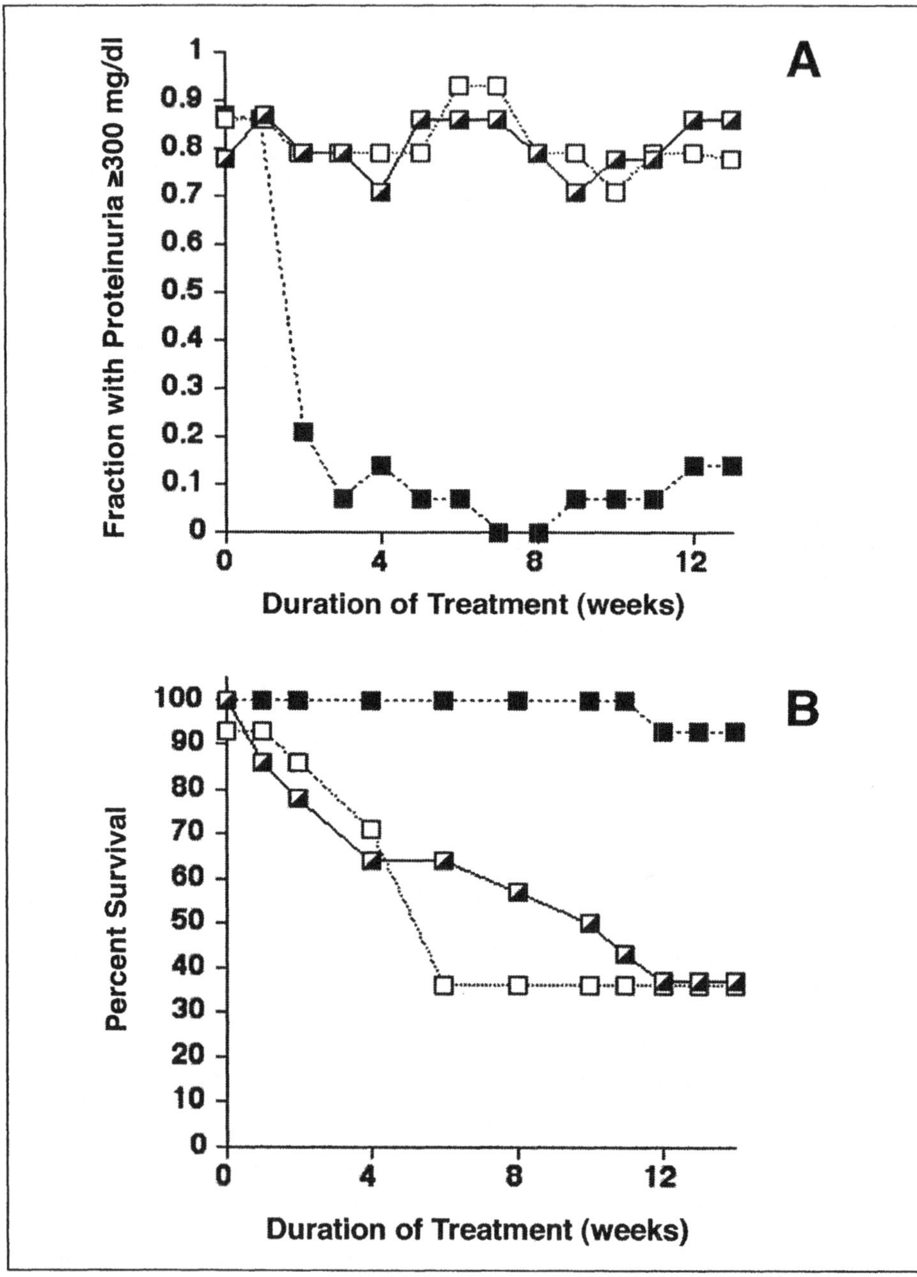

Figure 2. A. Development of proteinuria among mice with advanced renal disease during treatment with cyclophosphamide (half-solid boxes), CTLA4Ig (open boxes), or combined cyclophosphamide and CTLA4Ig (solid boxes). B. Percent Survival among mice with advanced renal disease during treatment with cyclophosphamide (half-solid boxes), CTLA4Ig (open boxes), or combined cyclophosphamide and CTLA4Ig (solid boxes). Reprinted with permission from reference 46: Daikh, D.I. and D. Wofsy, Cutting edge: Reversal of murine lupus nephritis with CTLA4Ig and cyclophosphamide. J Immunology, 2001. 166: p. 2913-6. Copyright 2001. The American Association of Immunologists.

References

1. Bretscher PA, Cohn M. Minimal model for the mechanism of antibody induction and paralysis by antigen. Nature 1968; 220(166):444-8.
2. Ledbetter JA, Imboden JB, Schieven GL et al. CD28 ligation in T-cell activation: evidence for two signal transduction pathways. Blood 1990; 75(7):1531-9.
3. Linsley PS, Brady W, Grosmaire L et al. Binding of the B cell activation antigen B7 to CD28 costimulates T cell proliferation and interleukin 2 mRNA accumulation. J Exp Med 1991; 173(3):721-30.
4. Bluestone JA. New perspectives of CD28-B7-mediated T cell costimulation. Immunity 1995; 2(6):555-9.
5. Linsley PS, Clark EA, Ledbetter JA. T-cell antigen CD28 mediates adhesion with B cells by interacting with activation antigen B7/BB-1. Proc Natl Acad Sci USA 1990; 87(13):5031-5.
6. Freeman GJ, Borriello F, Hodes RJ et al. Murine B7-1, an alternative CTLA4 counter-receptor that costimulates T cell proliferation and interleukin 2 production. J Exp Med 1993; 178(6):2185-92.
7. Green JM, Noel PJ, Sperling AI et al. Absence of B7-dependent responses in CD28-deficient mice. Immunity 1994; 1(6):501-8.
8. Hutloff A, Dittrich AM, Beier KC et al. ICOS is an inducible T-cell co-stimulator structurally and functionally related to CD28. Nature 1999; 397(6716):263-6.
9. Mages HW, Hutloff A, Heuck C et al. Molecular cloning and characterization of murine ICOS and identification of B7h as ICOS ligand. Eur J Immunol 2000; 30(4):1040-7.
10. Wang S, Zhu G, Chapoval AI et al. Costimulation of T cells by B7-H2, a B7-like molecule that binds ICOS. Blood 2000; 96(8):2808-13.
11. McAdam AJ, Chang TT, Lumelsky AE et al. Mouse inducible costimulatory molecule (ICOS) expression is enhanced by CD28 costimulation and regulates differentiation of CD4+ T cells. J Immunol 2000; 165(9):5035-40.
12. Gonzalo JA, Delaney T, Corcoran J et al. Cutting edge: the related molecules CD28 and inducible costimulator deliver both unique and complementary signals required for optimal T cell activation. J Immunol 2001; 166(1):1-5.
13. Brunet JF, Denizot F, Luciani MF et al. A new member of the immunoglobulin superfamily—CTLA-4. Nature 1987; 328(6127):267-70.
14. Krummel MF, Allison JP. CD28 and CTLA-4 have opposing effects on the response of T cells to stimulation. J Exp Med 1995; 182(2):459-65.
15. Walunas TL, Lenschow DJ, Bakker CY et al. CTLA-4 can function as a negative regulator of T cell activation. Immunity 1994; 1(5):405-13.
16. Gribben JG, Freeman GJ, Boussiotis VA et al. CTLA4 mediates antigen-specific apoptosis of human T cells. Proc Natl Acad Sci USA 1995; 92(3):811-5.
17. Leach DR, Krummel MF, Allison JP. Enhancement of antitumor immunity by CTLA-4 blockade. Science 1996; 271(5256):1734-6.
18. Waterhouse P, Penninger JM, Timms E et al. Lymphoproliferative disorders with early lethality in mice deficient in Ctla-4. Science 1995; 270(5238):985-8.
19. Perez VL, Van Parijs L, Biuckians A et al. Induction of peripheral T cell tolerance in vivo requires CTLA-4 engagement. Immunity 1997; 6(4):411-7.
20. Luhder F, Hoglund P, Allison JP et al. Cytotoxic T lymphocyte-associated antigen 4 (CTLA-4) regulates the unfolding of autoimmune diabetes. J Exp Med 1998; 187(3):427-32.
21. Jenkins MK, Mueller D, Schwartz RH et al. Induction and maintenance of anergy in mature T cells. Adv Exp Med Biol 1991; 292(1):167-76.
22. Tan P, Anasetti C, Hansen JA et al. Induction of alloantigen-specific hyporesponsiveness in human T lymphocytes by blocking interaction of CD28 with its natural ligand B7/BB1. J Exp Med 1993; 177(1):165-73.
23. Lenschow DJ, Zeng Y, Thistlethwaite JR et al. Long-term survival of xenogeneic pancreatic islet grafts induced by CTLA4lg. Science 1992; 257(5071):789-92.
24. Turka LA, Linsley PS, Lin H et al. T-cell activation by the CD28 ligand B7 is required for cardiac allograft rejection in vivo. Proc Natl Acad Sci USA 1992; 89(22):11102-5.
25. Milich DR, Linsley PS, Hughes JL et al. Soluble CTLA-4 can suppress autoantibody production and elicit long term unresponsiveness in a novel transgenic model. J Immunol 1994; 153(1):429-35.
26. Steinberg AD, Raveché ES, Laskin CA et al. NIH conference. Systemic lupus erythematosus: insights from animal models. Annals of Internal Medicine 1984; 100(5):714-27.

27. Izui S, McConahey PJ, Dixon FJ. Increased spontaneous polyclonal activation of B lymphocytes in mice with spontaneous autoimmune disease. Journal of Immunology 1978; 121(6):2213-9.
28. Klinman DM, Steinberg AD. Systemic autoimmune disease arises from polyclonal B cell activation. Journal of Experimental Medicine 1987; 165(6):1755-60.
29. Blaese RM, Grayson J, Steinberg AD. Increased immunoglobulin-secreting cells in the blood of patients with active systemic lupus erythematosus. American Journal of Medicine 1980; 69(3):345-50.
30. Frank MM, Hamburger MI, Lawley TJ et al. Fc-receptor dysfunction in lupus [letter]. New England Journal of Medicine 1979; 300(26):1487.
31. Cohen PL, Litvin DA, Winfield JB. Association between endogenously activated T cells and immunoglobulin-secreting B cells in patients with active systemic lupus erythematosus. Arthritis and Rheumatism 1982; 25(2):168-73.
32. Wofsy D, Seaman WE. Successful treatment of autoimmunity in NZB/NZW F1 mice with monoclonal antibody to L3T4. J Exp Med 1985; 161(2):378-91.
33. Finck BK, Linsley PS, Wofsy D. Treatment of murine lupus with CTLA4Ig. Science 1994; 265(5176):1225-7.
34. Nakajima A, Azuma M, Kodera S et al. Preferential dependence of autoantibody production in murine lupus on CD86 costimulatory molecule. Eur J Immunol 1995; 25(11):3060-9.
35. Liang B, Gee RJ, Kashgarian MJ et al. B7 costimulation in the development of lupus: autoimmunity arises either in the absence of B7.1/B7.2 or in the presence of anti-b7.1/B7.2 blocking antibodies. J Immunol 1999; 163(4):2322-9.
36. Kaneko H, Saito K, Hashimoto H et al. Preferential elimination of CD28+ T cells in systemic lupus erythematosus (SLE) and the relation with activation-induced apoptosis. Clin Exp Immunol 1996; 106(2):218-29.
37. Garcia-Cozar FJ, Molina IJ, Cuadrado MJ et al. Defective B7 expression on antigen-presenting cells underlying T cell activation abnormalities in systemic lupus erythematosus (SLE) patients. Clin Exp Immunol 1996; 104(1):72-9.
38. Liu MF, Liu HS, Wang CR et al. Expression of CTLA-4 molecule in peripheral blood T lymphocytes from patients with systemic lupus erythematosus. J Clin Immunol 1998; 18(6):392-8.
39. Heward J, Gordon C, Allahabadia A et al. The A-G polymorphism in exon 1 of the CTLA-4 gene is not associated with systemic lupus erythematosus. Ann Rheum Dis 1999; 58(3):193-5.
40. Heward JM, Allahabadia A, Carr-Smith J et al. No evidence for allelic association of a human CTLA-4 promoter polymorphism with autoimmune thyroid disease in either population-based case-control or family-based studies. Clin Endocrinol (Oxf) 1998; 49(3):331-4.
41. Mehrian R, Quismorio FP Jr, Strassmann G et al. Synergistic effect between IL-10 and bcl-2 genotypes in determining susceptibility to systemic lupus erythematosus. Arthritis Rheum 1998; 41(4):596-602.
42. Matsushita M, Tsuchiya N, Shiota M et al. Lack of a strong association of CTLA-4 exon 1 polymorphism with the susceptibility to rheumatoid arthritis and systemic lupus erythematosus in Japanese: an association study using a novel variation screening method. Tissue Antigens 1999; 54(6):578-84.
43. Allen RC, Armitage RJ, Conley ME et al. CD40 ligand gene defects responsible for X-linked hyper-IgM syndrome. Science 1993; 259(5097):990-3.
44. Aruffo A, Farrington M, Hollenbaugh D et al. The CD40 ligand, gp39, is defective in activated T cells from patients with X-linked hyper-IgM syndrome. Cell 1993; 72(2):291-300.
45. Durie FH, Foy TM, Masters SR et al. The role of CD40 in the regulation of humoral and cell-mediated immunity. Immunol Today 1994; 15(9):406-11.
46. Mohan C, Shi Y, Laman JD et al. Interaction between CD40 and its ligand gp39 in the development of murine lupus nephritis. J Immunol 1995; 154(3):1470-80.
47. Kalled SL, Cutler AH, Datta SK et al. Anti-CD40 ligand antibody treatment of SNF1 mice with established nephritis: preservation of kidney function. J Immunol 1998; 160(5):2158-65.
48. Daikh DI, Finck BK, Linsley PS et al. Long-term inhibition of murine lupus by brief simultaneous blockade of the B7/CD28 and CD40/gp39 costimulation pathways. J Immunol 1997; 159(7):3104-8.
49. Boumpas DT, Austin HAd, Vaughn EM et al. Controlled trial of pulse methylprednisolone versus two regimens of pulse cyclophosphamide in severe lupus nephritis. Lancet 1992; 340(8822):741-5.
50. Russell PJ, Hicks JD, Burnet FM. Cyclophosphamide treatment of kidney disease in (NZB x NZW) F1 mice. Lancet 1966; 1(7450):1280-4.

51. Russell PJ, Hicks JD. Cyclophosphamide treatment of renal disease in (NZB x NZW) F1 hybrid mice. Lancet 1968; 1(7540):440-1.
52. Daikh DI, Wofsy D. Cutting edge: reversal of murine lupus nephritis with CTLA4Ig and cyclophosphamide. J Immunol 2001; 166(5):2913-6.

Chapter 4

CTLA-4 in Rheumatoid Arthritis

Peter P. Sfikakis and Stamatis-Nick Liossis

Introduction

Rheumatoid arthritis (RA) is a chronic, systemic inflammatory disease characterised by symmetric polyarthritis of the small joints of the hands and feet and the larger appendicular joints. The etiology of RA is still unknown. Although several features of autoimmunity are prominent in these patients, the nature of the antigen(s) driving joint inflammation remains unclear. While the initiation phase of RA might result from an (auto)antigen-specific T cell response, the perpetuation of inflammation leading to joint destruction in the late chronic phases of the disease is a consequence of complex pathogenetic mechanisms involving aberrant interactions between T cells, macrophages, and synovial fibroblasts. T cells are the most frequently observed inflammatory cell in the rheumatoid joint (1-3). Not all authorities agree that RA is a purely T cell-mediated disease; however, an important pathogenetic role of the T cells has been clearly established. Besides their frequency, the fact that the majority possess both a memory and an activated T cell phenotype, as well as the association of RA with specific molecules of the major histocompatibility complex (MHC) class II underscore the central role of T cells in the pathogenesis of RA.[3,4]

Two signals are necessary for a productive T cell stimulation which is required for all T cell-dependent immune processes. The first signal confers specificity and is transduced via the T cell-surface antigen receptor (TCR) when antigen is properly presented to the T cell by an antigen presenting cell (APC) in the context of MHC molecules. The very same APC has to provide a second or co-stimulatory signal in order to sustain and enhance T cell activity; otherwise a state of long-term specific unresponsiveness or anergy is reached. While costimulation can be given by the interaction of several pairs of surface molecules between the T cell and the APC, the most important costimulatory signal is produced by the ligation of the CD28 molecule present on the surface of T cells to the B7 family of molecules [B7-1 (CD80) and/or B7-2 (CD86)] found on the surface of the professional APC.[5,6]

The CD28 molecule and the cytotoxic T lymphocyte-associated antigen 4 (CTLA-4 or CD152) have important similarities and perhaps more important differences. They are homologous proteins (31% sequence homology) belonging to the immunoglobulin superfamily. CTLA-4 shares the same counter-receptors with CD28 but has up to 100 times higher affinity to CD80 and CD86 compared to CD28.[6] While CD28 is expressed on the surface of virtually all the resting $CD4^+$ T cells and on the surface of 50% of the resting $CD8^+$ population, CTLA-4 can be found on the surface of activated T cells only. Although some reports have suggested that CTLA-4 has a function similar to that of CD28

CTLA-4 in Autoimmune Disease, edited by Flemming Pociot. ©2004 Eurekah.com.

as a secondary costimulator, most have argued for an opposing role of this molecule. In fact, cross-linking of CTLA-4 during activation of either Th1 or Th2 cells reduces cytokine production and arrests T cells in G1 phase.[7] Therefore, CTLA-4 can function by delivering a negative signal to the activated T cell that is thought to play an essential role in both the establishment of peripheral self-tolerence and the termination of a specific ongoing immune response.[8,9] These properties of CTLA-4 (presence on activated T cells only and inhibiting of the immune response) made it a plausible candidate for investigations in the context of autoimmune mediated diseases, including RA.[10]

This chapter will give an overview of the recently accumulating evidence implicating CTLA-4 in the pathogenesis of RA. Genetic studies have recently suggested that CTLA-4 polymorhisms are present in patients with RA, thus conferring susceptibility to the disease. Also, several reports discussed herein have shown altered expression of CTLA-4 and its counter-receptors, CD80 and CD86, in patients with RA contributing perhaps to the ongoing immune-mediated joint damage. Finally, effective blockade of the CD80/CD86:CD28/CTLA-4 pathway using the CTLA-4-immunoglobulin (CTLA4-Ig) construct was shown to prevent and/or ameliorate established disease in animal models of RA, indicating that this approach has the potential to lead to more effective strategies for the treatment of RA in the near future.

CTLA-4 Polymorphism in Patients with RA

During its earliest days, CTLA-4 function was a matter of debate. Most authorities however agreed, that CTLA-4 engagement delivers a negative, inhibitory signal to the activated T cell. This was firmly established with the study of CTLA-4-knockout mice that developed fatal massive lymphoproliferation.[11] It was therefore hypothesized that mutated or decreased-affinity polymorphs of CTLA-4, mapping at position 2q33, might contribute to immune-mediated diseases such as RA where an ongoing (auto)immune process is involved in the perpetuation of joint inflammation.

Within the past two years several genetic studies employing different methodologies have been performed in patients with RA with different ethnic backgrounds, reporting different results. Much attention has been drawn to the position +49 polymorphisms of the first exon of CTLA-4 and practically all the reports investigate this interesting area. It is further interesting because single nucleotide polymorphisms at position +49 of the CTLA-4 gene have been associated with diverse yet chronic inflammatory autoimmune conditions such as celiac disease, Grave's disease, multiple sclerosis, and type I diabetes (see Chapters 5, 6, 8, and 10, respectively). This polymorphism (49 A/G) results in codon 17 dimorphism producing an aminoacid exchange (Thr/Ala).

One study analyzed 258 Caucasian patients with RA and 456 controls. The authors propose that patients with RA are less frequently homozygotes for the Thr-17 substitution (32% versus 39% in the controls) but the Thr/Ala heterozygotes are more frequent compared to the controls (54% versus 46%). Analysis with respect to HLA-DRB1*04 revealed significantly more Ala in the homozygous or heterozygous state. Ala/Ala was particularly increased in RA patients carrying the HLA-DRB1*0401 subtype (12). Another study analyzed 138 patients with RA from Spain and 305 ethnically matched controls. The 49 A/G polymorphism of CTLA-4 exon 1 was analyzed along with position –318(C/T) of the CTLA-4 promoter. While no association was found for the –318 site, A/G heterozygosity at position +49 was over-represented among female patients with RA (48.5% vs. 33.8% in controls, odds ratio=2.0). It was the DR3+ patient group that demonstrated significant association with the 49A/G heterozygosity.[13]

However, the association between RA and CTLA-4 exon 1 polymorphism was not found in a study analyzing Japanese patients with RA. A novel screening method, the polymerase chain reaction-preferential homoduplex formation assay, was employed. Results showed that 44% of the Japanese patients with RA and systemic lupus erythematosus carry the CTLA-4 49 A/G polymorphism compared to 37% in the control population, a difference that was not significant. When analysis was performed with respect to HLA-DRB1 alleles, a very weakly significant association between RA and the CTLA-4 49G allele in HLA-DRB1*0405-positive patients was revealed, suggesting that the contribution of this CTLA-4 polymorphism was minor.[14] In contrast to these findings, an additional study asking the same question in 85 Japanese patients and 200 controls employed polymerase chain reaction-restriction fragment length polymorphism assay to study CTLA-4 exon 1 49A/G polymorphisms. The authors conclude that the AG genotype was found more frequently in patients with RA (59%) than in controls (44%). The association with RA was restricted to patients carrying the HLA-DRB1*0405 disease susceptibility allele.[15] Using the same methodology however, a study of CTLA-4 exon 1 49A/G dimorphism in patients from the UK (n=192) and from Spain (n=136) who were jointly analyzed, along with their appropriate ethnic-matched controls (n=96 and 144, respectively)again reported that no significant differences were found in the frequency of G allele or GG homozygosity in either the UK or the Spanish RA patients concluding that there was no association between RA and CTLA-4.[16]

A study analyzing RA families employed whole genome scan of 114 European sib pairs using 304 microsatellites. Regions outside the MHC locus (known to confer at least one-third of the genetic susceptibility to RA) were scanned with special emphasis. Additional loci were analyzed using additional families. Apart from the MHC locus, 14 other regions had nominal but not significant linkage. Interestingly, the one additional locus that had significant linkage with RA maps in chromosome 3 and although it is not CTLA-4, there is a strong possibility that it involves the CD80 and CD86 counter-receptors (17). Finally, the latest genetic study of them all further clarifies the issue of CTLA-4 exon 1 49 A/G polymorphism. Analyzing a genetically homogeneous Scandinavian population the investigators suggest that the 49 A allele is only associated with celiac disease and not RA or other chronic inflammatory autoimmune diseases. They suggest also that the +49A/G dimorphic alleles of CTLA-4 are in linkage disequilibrium with two distinct disease predisposing alleles with separate effects.[18]

The above studies present conflicting results regarding the association of CTLA-4 polymorphisms and RA. Moreover, none of these studies investigated whether CTLA-4 inhibitory function is impaired in T cells from RA patients carrying the CTLA-4 gene polymorhisms, as was indeed recently shown in patients with Graves' disease with CTLA-4 exon 1 polymorphisms.[19] Most importantly, three studies do not support an association between RA and such polymorphisms.[14,16,18] It is clear that the genetic contribution to the development of RA is far from simple; nevertheless it is also understood that if there was such an association with one genetic locus, all studies should agree on that finding.[20] Of note, while the gene encoding CD28 molecule is also mapping at chromosome 2q33, completion of the genomic sequence of this locus disclosed that a third costimulatory receptor, the inducible costimulator (ICOS) discovered last year, is so close to CTLA-4 gene on the chromosome that both are well within the limits of current genetic studies. Given that ICOS is necessary for normal antibody responses and may be protective against autoimmunity, perhaps we do not really know yet which the autoimmune disease-related candidate gene at chromosome 2q33 is.[21]

Increased Expression of CTL-4 on RA Synovial T Cells

Upregulation of CTLA-4 mRNA and expression of CTLA-4 on the T cell surface requires ligation of CD28 following a TCR-mediated signal. Although the regulation of CTLA-4 expression in RA has not been examined so far, studies on peripheral $CD28^+$ T cells from patients with RA have found that in vitro proliferative responses to both CD28-independent and CD28-dependent stimulation (using stable transfectants expressing the CD80 molecule) are normal, suggesting that the CD28-mediated signaling pathway in peripheral blood T cells is not intrinsically impaired in patients with either active or inactive RA.[22] Phenotypic analysis of healthy persons with HLA-DRB1*0401 and DR1 alleles, that are associated with increased risk of RA, revealed significantly higher numbers of $CD28^-$ T cells in the peripheral blood, while individuals with HLA-DR2, which is significantly under-represented in RA, have significantly fewer $CD28^-$ T cells than the normal mean.[23] Accordingly, patients with active, untreated RA have decreased numbers of functionally intact peripheral blood $CD28^+CD8^+$ T cells, that are related to severe disease.[22] Decreased numbers of $CD28^+$ T cells in the peripheral blood may result from selective migration of these cells in the rheumatoid joint, since comparative study of both peripheral blood and synovial fluid T cells from patients with active RA shows that virtually all synovial fluid T cells, versus 91% $CD4^+$ and 46% $CD8^+$ in the peripheral blood, coexpress the CD28 molecule.[22] Others have also confirmed that the vast majority of T cells isolated from fresh rheumatoid synovial tissue and fluid express the CD28 molecule.[24-26]

Significant expression of CTLA-4 has not been demonstrated on the surface of circulating T cells from healthy individuals or patients with autoimmune diseases, including RA. Liu et al reported recently that only a small percentage (i.e.,less than 2%) of peripheral blood T cells expressed detectable CTLA-4 levels in some RA patients. However, 5.44% of synovial fluid T cells and an impressive 28.76% of synovial membrane T cells were positive for CTLA-4 (27). Verwilgen et al have also reported that a significant number of T cells (mean of 15%) isolated from the synovial fluid of patients with RA, aspirated for therapeutic purposes, display increased CTLA-4 expression. CTLA-4-expressing synovial fluid T cells are activated, larger than normal and belong to the $CD4^+$ as well as to the $CD8^+$ T cell subsets. Interestingly, CTLA-4^+ synovial fluid T cells stained in some cases also positive for CD80.[24] This is a finding of potential importance, since ligation of CD80 expressed on the surface of T cells with CTLA-4 molecule found on the cell surface of the same or neighboring T cells is required for the induction of unresponsiveness by costimulation-deficient antigen presentation.[28] Whether CTLA-4 upregulation in RA is a primary or a secondary event has not been fully elucidated yet.

Altered Expression of the CTLA-4 Counter-Receptors, CD80, and CD86, in Patients with RA

The mechanisms proposed to explain the inhibitory activity of CTLA-4 involve either functional antagonism of CD28 or the delivery of an inhibitory signal that antagonizes the TCR-initiated signal. Although experimental evidence indicates that the latter mechanism is sufficient to explain the inhibitory effects of CTLA-4,[29] it is evident that engagement of CTLA-4 depends on the availability of the CD80 and CD86 molecules. CD80 and CD86 display a restricted expression pattern on APCs and depending on the type of APC, their expression can be induced by various stimuli. CD86 is constitutively expressed on resting peripheral monocytes and dendritic cells; its expression is upregulated following activation. Expression of CD80 at substantial levels on these cells appears to be primarily induced following activation. Unlike monocytes and dendritic cells, expression of CD80 and CD86 on all other APCs, such as synovial cells or T cells which are also capable to process and present antigens under certain conditions, requires stimulation. Following APC stimulation, CD86 is expressed 24-48 hours earlier and at higher levels than CD80. Moreover, the same stimulus can independently regulate CD80 and CD86 expression on APCs (reviewed in 30).

Expression of CD86 is clearly upregulated in the RA joint.[25,31-33] In the study of Liu et al CD86-expressing cells with macrophage-like morphology surrounded by lymphocyte aggregates are observed in two thirds of patients with long-standing RA undergoing therapeutic knee surgery, but in none of the osteoarthritic or normal synovia studied in parallel. CD86-expressing cells are more abundant than CD80-expressing cells in RA synovium in these patients.[25] CD86 immunohistochemical expression was also detected in synovia from patients with early-onset RA, predominantly in the lining layer, in a pattern that corresponded to the presence of CD68-positive macrophages.[33] A mean of 6% (range 2-15%) of mononuclear cells derived from synovial fluid from chronic RA patients express CD86 by flow-cytometric analysis; 40% of this popupation co-express the monocyte marker CD14.[25] Another study showed that the percentage of synovial fluid dendritic cells and monocytes that express CD86 is variable among patients with RA, depending on treatment. All patients who were not taking disease-modifying drugs had at least 20%, whereas 5 out of 7 patients under treatment had 11% or fewer $CD86^+$ dendritic cells in the synovial fluid.[31] Therefore, numbers of $CD86^+$ cells in the rheumatoid synovial fluid are clearly influenced by disease status (early versus long-standing disease or active versus inactive disease) and treatment, accounting for differences between studies reporting higher[25,31] or relatively lower numbers.[32] Of note, sorted synovial fluid dendritic cells lacking CD86 expression are able to spontaneously upregulate CD86 in vitro within 24 hours, suggesting that numbers of CD86 expressing cells in the inflammatory joint environment may change depending on local conditions.[31]

Inadequate, probably, CD80 expression by non-B cell APCs in vivo has been reported in patients with RA.[31-35] Neither freshly isolated peripheral blood dendritic cells, nor monocytes, from patients with RA expressed CD80, whereas both rheumatoid synovial fluid dendritic cells and monocytes expressed very low levels.[32] Minimal CD80 expression on rheumatoid synovial fluid dendritic cells[33] and synovial tissue[34] may be an additional factor contributing to a functional deficiency of, at least some, APCs demonstrated previously in the rheumatoid joint.[36] Moreover, expression of CD80 by non-professional APCs such as the fibroblast-like synoviocytes is lacking.[31,35] While fibroblast-like synoviocytes are $HLA\text{-}DR^+$ they do not express CD80 molecules. This may result in T cell anergy within the RA joint, since the allogeneic response of T cells to fibroblast-like synoviocytes could be restored by exogenously derived CD80 expressed on the surface of other cells and could be completely inhibited by CTLA4-Ig.[35] However, according to another study the LeuM3+ synovial cells express functional CD80 and CD86 molecules that are able to induce CD28 responsive nuclear transcription factor in synovial infiltrating CD28+ T cells. Such irradiated synovium-infiltrating T cells induced IL-1b, IL-6 and MMP-3 production by synovial cells. The authors propose that the production of the above cytokines and MMP-3 was mediated via CD80/CD86:CD28 interactions, since the addition of CTLA-4-Ig, or anti-CD80, or anti-CD86 monoclonal antibodies could abolish the production of the proinflammatory cytokines and MMP-3.[26]

On the other hand increased CD80 molecule expression has been surprisingly found on rheumatoid synovial T cells by both immunohistochemistry (24) and flow cytometry.[24,31] Although highly variable between patients, a mean of 20% of synovial fluid pre-activated, HLA-DR$^+$ T cells coexpress CD80. Moreover, while only 1% of osteoarthritic or normal synovial membrane T cells express CD80, up to 30% of T cells in synovial membrane samples from patients with chronic RA are CD80$^+$.[24] Finally, patients with active, early-stage RA not only have increased expression of CD80 on synovial fluid T cells, but also have significant numbers (mean of 4 %) of T cells expressing CD80 in the peripheral blood.[31] As previously noted, CTLA-4 ligation by CD80 on T cells is required for the induction of unresponsiveness by costimulation-deficient antigen presentation.[28] Taken together with the significant CTLA-4 expression on synovial fluid T cells, these findings reflect the increased activation state of T cells in patients with RA and may have important implications for the pathogenesis of the disease.

Potential Pathogenetic Role of CTLA-4 in T Cell-Mediated Interactions in RA

The possible role of CTLA-4 in the initiation phase of rheumatoid inflammation has not been addressed so far. Whether polymorphic CTLA-4 alleles are associated with impaired CTLA-4 molecule inhibitory function remains to be elucidated. However, several data presented above implicate CTLA-4 in the aberrant T-cell dependent mechanisms operating in the perpetuation of joint inflammation in patients with RA.

What is the functional significance of increased CTLA-4 expression in the RA joint? Is it just a normal T cell activation marker? As previously reported, the activation status of synovial T cells in RA is paradoxical. RA synovial fluid T cells express the activation markers CD69, HLA-DR and VLA-1 but just a few express the activation marker CD25 that represents the IL-2 receptor. This unusual set of activation T cell surface markers cannot be explained by single or even multiple rounds of activation in situ. It has been proposed that their phenotype may be explained either by the recruitment of pre-activated T cells in the synovium, or by the acquisition of activation markers during their migration to the joint and their contact with the endothelium.[37] Is CTLA-4 found in RA T cells per se defective? The functional capacity of CTLA-4 in RA has been explored rather indirectly using in vitro synovial mononuclear cell culture experiments. Blocking of CTLA-4 with an anti-CTLA-4 monoclonal antibody resulted in a dose-dependent increase in the production of the proinflammatory cytokines TNF-α and IL-1β suggesting that it is functionally intact.[27] This observation would further suggest that upregulation of CTLA-4 on RA synovial T cells might have a protective role in joint inflammation in these patients. On the other hand, addition of anti-CTLA-4 antibodies in a primary allogeneic mixed lymphocyte culture using synovial fluid T cells from patients with RA as stimulator cells did not enhance T cell stimulating capacity but resulted in a 33% inhibition of proliferation.[24] Therefore, if CTLA-4 inhibitory function is intact, the question remains as to why blockade of increased CTLA-4 expression on synovial T cells do not enhance their activation status.

As recently reported, the CTLA4-initiated T cell inhibitory signal is delivered via the TCRζ chain. When CTLA-4 associates with TCRζ the TCR-triggered tyrosine phosphorylation cascade and hence the TCR-initiated signaling is inhibited.[38] In contrast to systemic lupus erythematosus where TCRζ chain is lacking from the peripheral blood T cells in the majority of patients,[39] it is the synovial T cells but not the circulating ones that display ζ chain deficiency in RA.[40] It can be thus assumed that CTLA-4 is present in increased amounts on the surface of synovial T cells in patients with RA, yet it cannot deliver its inhibitory signal because an important mediator of this pathway (TCRζ) is decreased or even missing from the synovial T cell population. Clearly, further studies are needed to clarify the particular biochemical basis for the abnormal functional outcome of CTLA-4 ligation, if any, in RA.

Blocking autoantibodies may also contribute to a decreased CTLA-4 function in RA. Indeed, a recent study reported that in the sera of 18.8% of patients with RA, autoantibodies to CTLA-4 were detected. Such autoantibodies were not disease-specific since they were also found in significant percentages of patients with other autoimmune rheumatic diseases such as systemic lupus erythematosus and, in particular, Behçet's disease where their presence protected from the development of uveitis.[41] Anti-CTLA-4 antibodies may thus bind on CTLA-4 molecule expressed on the surface of T cells and may modulate the in vivo immune response.

One might also speculate that it is the function of the CTLA-4 counter-receptors, CD80 and CD86 that is deficient. Nevertheless, there are several reports using the CTLA4-Ig construct as a probe to explore CD80/CD86 functional status which appears to be intact.[4,26,32,33,35] It is thus reasonable to assume that the functional integrity of CTLA-4 when viewed solely as a counter-receptor for CD80/CD86 molecules is not affected in the RA joint environment. But the expansion of $CD28^{-}$ T cells encountered in RA and the CD28-independent costimulation these T cells need in order to expand, may propose a relatively secondary role of pathways mediated by CD80/CD86:CD28/CTLA-4 intreractions in the pathogenesis of RA.[42] As discussed earlier though, CTLA-4 may have a T cell inhibitory role for the same T cell that bears CTLA-4 on its cell surface.

RA as a Potential Target of CTLA-4-Related Therapeutic Interventions—Future Directions

As the two-signal model predicts, blocking costimulatory signals leads to defective T cell responses and development of T cell anergy or tolerance. Taking advantage of the very high affinity of CTLA-4 to the B7 molecules, the CTLA4-Ig construct has been used extensively to block the CD80/CD86:CD28/CTLA-4 interactions in vivo. The CTLA4-Ig chimeric protein construct consists of the extracellular domain of CTLA-4 and the hinge-C_H2-C_H3 region of IgG1. In vitro studies have previously shown that CTLA4-Ig is a potent inhibitor of T cell activation, inducing also T cell hyporesponsiveness or anergy to subsequent antigen challenge.[30]

In an experimental model of RA, the collagen-induced arthritis in the BB rat, the use of CTLA4-Ig construct prevented the onset of arthritis. Treatment with CTLA4-Ig had to begin before the immunizations with the bovine type II collagen, resulting in decreased production of antibodies to bovine collagen and in normal joint histology. However, several days after the treatment with CTLA4-Ig was terminated, delayed type reactions to bovine collagen were not avoided and resulted in a similarly aggressive and destructive joint disease irrespective of weather the animal had been pre-treated or not with CTLA4-Ig. Nevertheless, this molecule was not used following the onset and establishment of the disease.[43] When CTLA4-Ig was given at the same time the immunizations took place, the development of arthritis was prevented. When treatment was delayed until after the onset of clinical arthritis, it resulted in amelioration of the disease.[44] To bypass the need for repeated injections and at the same time to ensure that enough quantities of CTLA4-Ig are constantly delivered to the joints, adenovirus-mediated gene transfer of CTLA4-Ig in established murine collagen-induced arthritis has been also employed. The recombinant adenovirus encoding CTLA4-Ig was injected intravenously once in DBA/1 mice at the onset of arthritis, while other groups of arthritic mice were given CTLA4-Ig or an anti-CTLA-4 monoclonal antibody. CTLA4-Ig, given either as the recombinant protein or as the adenovirally encoded protein were equally successful in ameliorating the disease, while the group of mice receiving anti-CTLA-4 monoclonal antibody did not show any disease improvement. Clinical parameters as well as both cellular and humoral immunity markers of the disease and IFN-γ production were significantly suppressed by CTLA4-Ig treatment.[45] Moreover, using CTLA4-Ig in another systemic lupus-like autoimmunity animal model, the production of the commonest autoantibody of RA, rheumatoid factor, was blocked.[46]

Abrams and colleagues have recently studied the in vivo effects of CD80/CD86:CD28/CTLA-4 blockade in patients with psoriasis vulgaris, a T cell-mediated disease which is often accompanied by inflammatory arthritis. Infusions of soluble chimeric protein CTLA4-Ig were well tolerated and resulted in a 50% or greater sustained improvement in clinical disease activity in half of the patients.[47] Administration of CTLA4-Ig caused a marked reduction of cellular activation of lesional T cells, keratinocytes, dendritic cells and vascular endothelium that was associated with the clinical improvement in these patients.[48] Taken together with the proven beneficial effects of CTLA4-Ig treatment in the animal models of RA, it is reasonable to predict that human RA would be the next target of such a therapeutic approach.[49]

Finally, a novel strategy to enhance CTLA-4-mediated inhibition of TCR signaling employing a surface-linked single-chain antibody to CTLA-4 was recently reported. When a specific, membrane bound, single chain, anti-CTLA-4 monoclonal antibody was expressed on the surface of an APC, together with the TCR ligand, the corresponding T cell response was markedly attenuated. This may be explained because under this in vitro setting, CTLA-4 molecules were efficiently co-crosslinked to TCR without being bound to its normal costimulatory counter-receptors.[50] Such an approach would also represent a way to carefully manipulate CTLA-4 function and may have an important impact in the future treatment of human autoimmune diseases, including patients with RA.

References

1. Sfikakis PP, Tsokos GC. Lymphocyte adhesion molecules in autoimmune rheumatic diseases: basic issues and clinical expectations. Clin Exper Rheumatol 1995; 13:763-777.
2. Kingsley G, Panayi GS. Joint destruction in rheumatoid arthritis: biological bases. Clin Exper Rheumatol 1997; 15(suppl 17):S3-14.
3. Pope RM, Hohlfeld R. Pathogenesis of Rheumatoid arthritis. In: Tsokos GC, ed. Principles of Molecular Rheumatology. Totowa: Humana Press, 2000:325-362.
4. Panayi GS. T-cell dependent pathways in rheumatoid arthritis. Curr Opin Rheumatol 1997; 9:236-240
5. Lenschhow DJ, Walunas TL, Bluestone JA. CD28/B7 system of T cell costimulation. Ann Rev Immunol 1996; 14,233-258.
6. Reiser H, Stadecker J. Costimulatory B7 molecules in the pathogenesis of infectious and autoimmune diseases. N Engl J Med 1996; 335:1369-1376.
7. Alegre ML, Shiels H, Thompson CB, Gajewski TF. Expression and function of CTLA-4 in Th1 and Th2 cells. J Immunol 1998; 161:3347-3356.
8. Allison JP, Krummel MF. The Yin and the Yang of T cell costimulation. Science 1995; 270:932-933.
9. Bluestone JA. Is CTLA-4 a master switch for peripheral T cell tolerance? J Immunol 1997; 158:1989-1993.

10. Kristiansen OP, Larsen ZM, Pociot F. CTLA-4 in autoimmune diseases: a general susceptibility gene to autoimmunity? Genes Immun 2000; 1:170-184.
11. Tivol EA, Boriello F, Schweitzer AN et al. Loss of CTLA4 leads to massive lymphoproliferation and fatal multiorgan tissue destruction, revealing a critical negative regulatory role of CTLA-4. Immunity 1995; 3:541-547.
12. Seidl C, Donner H, Fischer B et al. CTLA4 codon 17 dimorphism in patients with rheumatoid arthritis. Tissue Antigens 1998; 51:62-66.
13. Gonzales-Escribano MF, Rodriguez R, Valenzuela A et al. CTLA4 polymorphisms in Spanish patients with rheumatoid arthritis. Tissue Antigens 1999; 53:296-300.
14. Matsushita M, Tsuchiya N, Shiota M et al. Lack of a strong association of CTLA-4 exon 1 polymorphism with the susceptibility to rheumatoid arthritis and systemic lupus erythematosus in Japanese: an association study using a novel variation screening method. Tissue Antigens 1999; 54:578-584.
15. Yanagawa T, Gomi K, Nakao EI et al. CTLA-4 gene polymorhisms in Japanese patients with rheumatoid arthritis. J Rheumatol 2000; 12:2740-2742
16. Barton A, Myerscough A, John S et al. A single nucleotide polymorphism in exon 1 of cytotoxic T-lymphocyte associated-4 (CTLA-4) is not associated with rheumatoid arthritis. Rheumatology (Oxford) 2000; 39:63-66.
17. Cornelis F, Faure S, Martinez M et al. New susceptibility locus for rheumatoid arthritis suggested by a genome-wide linkage scan. Proc Natl Acad Sci USA 1998; 95:10746-10750.
18. Naluai AT, Nilsson S, Samuelsson L et al. The CTLA4/CD28 gene region on chromosome 2q33 confers susceptibility to celiac disease in a way possibly distinct from that of type I diabetes and other chronic inflammatory disorders. Tissue Antigens 2000; 56:350-355.
19. Kouki T, Sawai Y, Gardine CA et al. CTLA-4 gene polymorphism at position 49 in exon 1 reduces the inhibitory function of CTLA-4 and contributes to the pathogenesis of Graves' disease. J Immunol 2000; 165:6606-6611.
20. Risch N. Searching for genes in complex diseases: lessons from systemic lupus erythematosus. J Clin Invest 2000; 105:1503-1506.
21. Linsley PS. T cell activation: you can't get good help. Nature Immunol 2001; 2:139-140.
22. Sfikakis PP, Zografou E, Viglis V et al. CD28 expression on T cell subsets in vivo and CD28-mediated T cell response in vitro in patients with rheumatoid arthritis. Arthritis Rheum 1995; 38:649-654.
23. Chapman A, Stewart SJ, Nepom GT et al. CD11b+CD28-CD4$^+$ human T cells. J Immunol 1996; 157:4771-4780.
24. Verwilghen J, Lovis R, De Boer M et al. Expression of functional B7 and CTLA4 on rheumatoid synovial T cells. J Immunol 1994; 153:1378-1385.
25. Liu MF, Kohsaka H, Sakurai H et al. The presence of costimulatory molecules CD86 and CD28 in rheumatoid arthritis synovium. Arthritis Rheum 1996; 39:110-114.
26. Shinoyama Y, Nagafuchi H, Suzuki N et al. Synovium infiltrating T cells induce excessive synovial cell function through CD28/B7 pathway in patients with rheumatoid arthritis. J Rheumatol 1999; 10:2094-2101
27. Liu MF, Yang CY, Li JS et al. Increased expression of down-regulatory CTLA-4 molecule on T lymphocytes from rheumatoid synovial compartment. Scand J Immunol 1999; 50:68-72.
28. Chai JG, Vendetti S, Amofah E et al. CD152 ligation by CD80 on T cells is required for the induction of unresponsiveness by costimulation-deficient antigen presentation. J Immunol 2000; 15:3037-3042.
29. Fallarino F, Fields PE, Gajewski TF. B7-1 engagement of cytotoxic T lymphocyte antigen 4 inhibits T cell activation in the absence of CD28. J Exp Med 1998; 188:205-210.
30. Boussiotis VA, Freeman GJ, Nadler LM. Adhesion and costimulatory molecules. In: Tsokos GC, ed. Principles of Molecular Rheumatology. Totowa: Humana Press, 2000:87-108.
31. Ranheim EA, Kipps TJ: Elevated expression of CD80 (B7/BB1) and other accessory molecules on synovial fluid mononuclear cell subsets in rheumatoid arthritis. Arthritis Rheum 1994; 37:1637-1646.
32. Thomas R, Quinn C. Functional differentiation of dendritic cells in rheumatoid synovium. J Immunol 1996; 156:3074-3086.
33. Summers KL, O'Donnell JL, Williams LA et al. Expression and function of CD80 and CD86 costimulator molecules on synovial dendritic cells in chronic arthritis. Arthritis Rheum 1996; 39:1287-1291.

34. Balsa A, Dixey J, Sansom DM et al. Differential expression of the costimulatory molecules B7.1 (CD80) and B7.2 (CD86) in rheumatoid synovial tissue. Br J Rheumatol 1996; 35:33-37.
35. Corrigall VM, Solau-Gervais E, Panayi GS. Lack of CD80 expression by fibroblast-like synoviocytes leading to anergy in T lymphocytes. Arthritis Rheum 2000; 43:1606-1615
36. Iglesias A, Deulofeut H, Dubey DP et al. Functional deficiency of antigen-presenting cells in the synovial fluid of rheumatoid arthritis. Hum Immunol 35:1992; 109-115.
37. Ioannone F, Corrigall VM, Kingsley GH et al. Evidence for the continuous recruitement and activation of T cells into the joints of patients with rheumatoid arthritis. Eur J Immunol 1994; 24:2706-2713.
38. Lee KM, Chuang E, Griffin M et al. Molecular basis of T cell inactivation by CTLA-4. Science 1998; 282:2263-2266.
39. Liossis SNC, Ding XZ, Dennis GJ et al. Altered pattern of TCR/CD3-mediated protein-tyrosyl phosphorylation in T cells from patients with systemic lupus erythematosus. J Clin Invest 1998; 101:1448-1457.
40. Berg L, Ronnelid J, Klareskog L et al. Down-regulation of the T cell receptor CD3 zeta chain in rheumatoid arthritis (RA) and its influence on T cell responsiveness. Clin Exp Immunol 2000:120:174-182.
41. Matsui T, Kurokawa M, Kobata T et al. Autoantibodies to T cell costimulatory molecules in systemic autoimmune diseases. J Immunol 1999; 162:4328-4335
42. Park W, Weyand CM, Schmidt D et al. Co-stimulatory pathways controlling activation and peripheral tolerance of human $CD4^{+}CD28^{-}$ T cells. Eur J Immunol 1997; 27:1082-1090.
43. Knoerzer DB, Karr RW, Schwartz BD et al. Collagen-induced arthritis in the BB rat. Prevention of disease by treatment with CTLA-4-Ig. J Clin Invest 1995; 96:987-993.
44. Webb LM, Walmsley MJ, Feldmann M. Prevention and amelioration of collagen-induced arthritis by blockade of the CD28 co-stimulatory pathway: requirement for both B7-1 and B7-2. Eur J Immunol 1996; 26:2320-2328.
45. Quatrocchi E, Dallman MJ, Feldmann M. Adenovirus-mediated gene transfer of CTLA-4Ig fusion protein in the suppression of experimental autoimmune arthritis. Arthritis Rheum 2000; 43:1688-1697.
46. Takiguchi M, Murakami M, Nakagawa I et al. Blockade of CD28/CTLA4-B7 pathway prevented the autoantibody-related diseases but not lung disease in MRL/lpr mice. Lab Invest 1999; 79:317-326.
47. Abrams JR, Lebwohl MG, Guzzo CA et al. CTLA4Ig-mediated blockade of T-cell costimulation in patients with psoriasis vulgaris. J Clin Invest 1999; 103:1243-1252.
48. Abrams JR, Kelley SL, Hayes E et al. Blockade of T lymphocyte costimulation with CTLA4Ig reverses the cellular pathology of psoriatic plaques, including the activation of keratinocytes, dendritic cells, and endothelial cells. J Exp Med 2000; 192:681-694.
49. Sfikakis PP, Via CS. Expression of CD28, CTLA4, CD80, and CD86 molecules in patients with autoimmune rheumatic diseases: Implications for immunotherapy. Clin Immunol Immunopathol 1997; 83:195-198.
50. Griffin MD, Hong DK, Holman PO et al. Blockade of T cell activation using a surface-linked single-chain antibody to CTLA-4 (CD152). J Immunol 2000; 164:4433-4442.

Chapter 5

CTLA-4 in Autoimmune Thyroid Disease and Vitiligo

E. Helen Kemp, Elizabeth A. Waterman and Anthony P. Weetman

Abbreviations: A, adenine; C, cytosine; CTLA-4, cytotoxic T lymphocyte antigen 4; G, guanine; HLA, human leukocyte antigen; ICAM-1, intracellular adhesion molecule 1; kDa, kilodalton; MHC, major histocompatibility complex; PPT, postpartum thyroiditis; T, thymine; TAO, thyroid-associated ophthalmopathy; TSH, thyroid-stimulating hormone; UTR, untranslated region.

Introduction

Autoimmune thyroid disease is frequent, affecting around 2% of the female population and approximately 0.2% of men. The two main forms are Graves' disease, the hallmark of which is hyperthyroidism caused by antibodies that stimulate the TSH receptor, and autoimmune hypothyroidism in which there is a destruction of thyroid tissue by cytotoxic T cells, cytokines and antibody-dependent mechanisms. Autoimmune thyroid disorders are often associated with other organ-specific autoimmune diseases, especially the endocrinopathies type 1 diabetes mellitus and Addison's disease, and with vitiligo. This latter skin disorder is therefore included in this chapter, the other conditions being dealt with elsewhere.

Like almost all autoimmune diseases, those affecting the thyroid have a multifactorial etiology comprising several known (and most likely more unknown) genetic determinants which interact with environmental factors including stress, iodide-intake and agents that disturb immunoregulatory pathways. The genetic component in autoimmune thyroid disease is demonstrated by the frequent appearance of similar disease in relatives, the concordance rate of around 20-30% in monozygotic twins, and the well-established association with HLA alleles, particularly the DR3 specificity in Caucasians.[1] Nontheless, the effect of genes appears to be relatively modest, and that of HLA even more so, indicating the influence of non-MHC-encoded genetic factors. Of the numerous candidates proposed in this regard, polymorphisms of the CTLA-4 gene are now the best established.

Many recent studies have been conducted to determine if associations exist between autoimmune thyroid disease and a number of different CTLA-4 polymorphic markers, including: an exon 1 A/G polymorphism at position 49; a promoter C/T polymorphism at position -318; a microsatellite $(AT)_n$ repeat polymorphism in the 3' untranslated region (UTR) of exon 4 at position 642. Below, the main autoimmune disorders of the thyroid and their associations with the various polymorphisms of the CTLA-4 gene will be considered.

CTLA-4 in Autoimmune Disease, edited by Flemming Pociot. ©2004 Eurekah.com.

Graves' Disease

Graves' disease is characterised by hyperthyroidism, a diffuse goiter and signs of ophthalmopathy and/or dermopathy.[2] The condition is caused by thyroid-stimulating antibodies which bind to and activate the thyrotropin receptor on thyroid cells.[2] Patients may also have high serum concentrations of antibodies against thyroglobulin and thyroid peroxidase.[2] Susceptibility to Graves' disease is influenced by a mixture of genetic factors, including certain HLA specificies,[1,2] and environmental determinants, the female sex being a particular risk element.[2]

Many reports have documented the analysis of the three known polymorphisms of CTLA-4, with respect to Graves' disease. The first description of an association was by Yanagawa et al in 1995,[3] and this concerned the exon 4 microsatellite $(AT)_n$ repeat polymorphism in Graves' disease patients. In most, but not all, subsequent studies, which are summarised in Table 1, this result has been confirmed, in both Caucasian and Asian populations. The linkage disequilibrium between the three main polymorphisms analysed has ensured that significant associations have been apparent whichever is chosen.

As well as case-control studies, family studies have been conducted but with less clear results. One family-based linkage analysis did report the CTLA-4 gene as a major risk factor for the development of Graves' disease.[4] However, in 48 multiplex families with 142 affected subjects, negative lod scores were obtained for CTLA-4 as a candidate gene in classical linkage analysis.[5] However, the family-based transmission disequilibrium test gave confirmatory results, with increased transmission of the position 49 G allele from heterozygous parents to the affected offspring compared with unaffected offspring.[6] This study also found a relationship between the severity of hyperthyroidism at diagnosis and the presence of the G allele. The differences between these two sets of results reflects the ability of the transmission disequilibrium test to pick up smaller genetic effects than classical linkage analysis, a property shared with case-control studies. No additional effect of CTLA-4 polymorphisms was found in patients who had both type 1 diabetes mellitus and Graves' disease compared with those with either disease alone.[7]

Thyroid-Associated Ophthalmopathy

Thyroid-associated ophthalmopathy (TAO) is an autoimmune disorder affecting the orbit. The disease has a close association with Graves' hyperthyroidism, being clinically evident in 25-50% of patients.[8] Major manifestions include periorbital edema, extraocular muscle dysfunction, exophthalmos and optic nerve compression which lead to visual impairment. Although the earliest pathological feature of Graves' ophthalmopathy is the appearance of T and B lymphocytes in the extraocular muscles, the target antigen of these immune cells in the orbit remains to be clearly identified. Age and gender can influence the severity of TAO with both advancing years and the male sex known to be risk factors.[8] In addition, the presence and severity of eye disease in Graves' disease patients appears to be associated with smoking ,[10] but no consistent association with HLA specificities has yet been demonstrated.

With respect to CTLA-4, a recent study examined the association of the biallelic polymorphism (A/G) at codon 17 of the CTLA-4 gene in 94 patients with TAO which were defined, according to the American Thyroid Association, as class 3 or above When the frequency of the G-carrying genotype in the TAO patient group was compared with.[11] either control subjects or with Graves' disease patients without ophthalmopthy, an

Table 1. Case-control studies of CTLA-4 gene polymorphisms in Graves' disease

CTLA-4 Polymorphism	Population	Results[1]	Reference
Microsatellite $(AT)_n$ repeat polymorphism in the 3'UTR of exon 4	USA	$P = 0.002$ for the 106 base pair allele	3
	UK	$P = 0.006$ for the 106 base pair allele	13
	Japanese	$P < 0.01$ for the 102 and 106 base pair alleles	20
Polymorphism A/G at position 49 of exon 1	German/ Canadian	$P < 0.00005$ for the G allele	57
	Japanese	$P < 0.01$ for the G allele	58
	UK	$P < 0.0002$ for the G allele	6
	Korean	$P = 0.04$ for the G allele	21
	Japanese	$P = 0.049$ for the G allele	19
Polymorphism C/T at position -318 of the CTLA-4 promoter	German/ Canadian	$P = 0.006$ for the C allele	23
	Korean	$P = 0.02$ for the C allele	21
	UK	$P = 0.29$ for the C allele	22
	Hong Kong Chinese	$P = 0.02$ for the C allele	22

[1] *P* value calculated from comparison of patients with control subjects. $P < 0.05$ are considered significant.

association was evident between the presence of the G allele and eye disease (Table 2).[11] Furthermore, the strength of the association of the G allele with ophthalmopathy was reported to increase with the severity of the disease.[11] Thus, the association of ophthalmopathy with the G-carrying genotype had an odds ratio of 1.5, 1.7 and 3.1 for mild, intermediate and severe grades of eye disease, respectively, when compared to Graves' disease patients with no eye signs. However, at least three other studies have found either no increase in the G allele or no difference in the distribution of the microsatellite $(AT)_n$ repeat alleles in TAO patients when compared with individuals with Graves' disease who do not have ophthalmopathy.[12-14] Further work is needed in this area to determine whether there is an additional risk of eye disease in Graves' disease patients with particular CTLA-4 polymorphisms. At present, it is our view that even if such a contribution does exist, it is modest, especially in comparison to the risk conferred by smoking.

Autoimmune Hypothyroidism

Autoimmune hypothyroidism is common with a prevalence of at least 1% in the female population.[15] The disease can be characterised by the presence of a goiter, as in Hashimoto's thyroiditis, or not, as in primary myxedema. Thyroid tissue injury in autoimmune hypothyroidism is likely to result from a combination of humoral and cell-mediated effector mechanisms. Thyroid peroxidase antibodies,[15] which are found in a majority of patients, can fix complement and participate in antibody-dependent cell-mediated cytotoxicity. Thyrocyte-specific cytotoxic T cells are probably also involved.

Table 2. Frequency of the CTLA-4 G allele of the A/G polymorphism in TAO patients and controls[1,2]

Genotype	TAO ($n = 94$)	Controls ($n = 90$)	Graves' Disease[3] ($n = 94$)	Odds Ratio[4]	*P* Value[4]	Odds Ratio[5]	*P* Value[5]
G allele	73 (78%)	45 (50%)	56 (60%)	3.48	0.00011	2.36	0.012

[1] *n*, number of individuals.
[2] Data from Vaidya et al [11].
[3] Graves' disease patients without ophthalmopathy.
[4] Results calculated from comparison with controls.
[5] Results calculated from comparison with Graves' disease patient group without ophthalmopathy.

In addtion to female sex, certain HLA specificities are associated with Hashimoto's thyroiditis and primary myxedma.[17]

Associations between autoimmune hypothyroidism and the microsatellite $(AT)_n$ repeat and exon 1 A/G polymorphisms of the CTLA-4 gene were initially reported in case-control studies using Caucasian subjects (Table 3).[13-18] This was later confirmed for a Japanese population.[19] In contrast to these studies, microsatellite allele frequencies were found not to significantly differ between Hashimoto's thyroiditis patients and controls in Japanese subjects (Table 3),[20] and no association was evident between the A/G polymorphism and Hashimoto's thyroiditis in Koreans (Table 3).[21] It is possible that the different genetic backgrounds of the patient groups might contribute to the different results obtained in these studies, as might variations in the number of subjects analysed. With respect to the C/T promoter polymorphism of CTLA-4, no significant association with autoimmune hypothyroidism has been reported for several distinct populations.[21-23]

Postpartum Thyroiditis

Postpartum thyroiditis (PPT) is a well defined autoimmune disorder affecting 4-6% of women during the year after delivery.[24] The condition is due to an unexplained exacerbation of pre-existing subclinical thyroid autoimmunity during the postpartum period. The disease is characterised by transient thyroid dysfunction comprising thyrotoxicosis and/or hypothyroidism although, in around 25% of women, permanent hypothyroidism can follow for several years after the apparent resolution of PPT.[25] The immunological features of PPT are similar to other forms of autoimmune thyroid disease including the presence of thyroid peroxidase autoantibodies,[26] abnormalities in the circulating T cell population[27] and a goiter with lymphocytic infiltration.[27] There are differences in prevalence of PPT in certain ethnic groups,[28] suggesting that the development of the disorder may be influenced by genetic and/or environmental factors. So far, however, only weak and inconsistent associations with various HLA specificities have been reported.[29]

A recent case-control study of the CTLA-4 microsatellite $(AT)_n$ repeat polymorphism in 122 postpartum women found no significant association of the 106 base pair allele and the incidence of PPT (Table 4).[30] This may reflect the heterogeneous nature of the disease itself and the numerous genetic susceptibility factors that probably contribute to the wide

Table 3. Case-control studies of CTLA-4 gene polymorphisms in autoimmune hypothyroidism

CTLA-4 Polymorphism	Population	Result[1]	Reference
Microsatellite $(AT)_n$ repeat polymorphism in the 3'UTR of exon 4	UK	$P = 0.02$ for the 106 base pair allele	13
	Japanese	$P < 0.10$ (but not significant) for the 102 and 106 base pair alleles	20
Polymorphism A/G at position 49 of exon 1	German/ Canadian	$P = 0.02$ for the G allele	18
	Japanese	$P = 0.029$ for the G allele	19
	Korean	$P = 0.37$ for the G allele	21

[1] P value calculated from comparison of patients with control subjects. $P < 0.05$ are considered significant.

clinical spectrum of PPT, making the effect of an individual locus difficult to ascertain. However, it is of interest to note that the group of women developing permanent hypothyroidism had the highest frequency of the 106 base pair $(AT)_n$ repeat allele, suggesting that these most resemble typical autoimmune hypothyroidism.

Vitiligo

In contrast to autoimmune thyroid disorders, there is a dearth of information concerning CTLA-4 polymorphisms in relation to vitiligo. Only two studies have been undertaken which have examined the association of vitiligo with the microsatellite $(AT)_n$ repeat alleles and the A/G polymorphism at position 49 of exon 1.

Vitiligo is an acquired idiopathic hypomelanotic disorder characterised by circumscribed depigmented macules resulting from the loss of cutaneous melanocytes. The cause of the disorder remains obscure, but an autoimmune aetiology has been suggested. The frequent association of vitiligo with autoimmune diseases[31,32] and studies demonstrating that most vitiligo patients have antibodies against melanocytes[33,34] support this hypothesis. Although the exact role played by these antibodies in the pathogenesis of vitiligo has not been determined, there is a correlation between their presence and level and the extent[35] and activity of the disease.[36] In addition, the sera from vitiligo patients can induce damage to human melanocytes in vitro by antibody-dependent cellular cytotoxicity,[37] and in vivo following passive immunisation of nude mice grafted with human skin.[38]

Autoantibodies detected in vitiligo patients are most commonly directed against pigment cell antigens with molecular weights of 35 kDa, 40-45 kDa, 75 kDa, 90 kDa and 150 kDa which are located on the surface of the cell[39,40] Although the proteins have not been specifically identified, some appear to be common tissue antigens while others are preferentially expressed on pigment cells.[39] In addition, antibodies to the melanocyte-specific proteins tyrosinase,[41-43] tyrosinase-related protein-1,[44] tyrosinase-related protein-2[45,46] and Pmel17[47] have been detected in the sera of patients with vitiligo.

Table 4. Frequency of the CTLA-4 106 base pair $(AT)_n$ microsatellite allele in postpartum women and in controls[1]

Group	Number of Patients	Allele Frequency	*P* Value[2]	Odds Ratio[2]
Postpartum women	122	57 (23.4%)	0.30	1.25
PPT+[3]	64	33(25.8%)	0.16	1.43
PPT-[4]	58	23 (19.8%)	1.00	1.02
Permanent hypothyroidism[5]	24	7 (29.1%)	0.13	1.70
Controls	161	63 (19.6%)		

[1]Data from Waterman et al [30].
[2]Compared to controls.
[3]PPT+, women with postpartum thyroid dysfunction.
[4]PPT-, women without postpartum thyroid dysfunction.
[5]Women who develop permanent hypothyroidism.

Evidence for the involvement of cellular immunity in the etiopathogenesis of vitiligo is provided by the presence of activated $CD4^+$ and $CD8^+$ cells at the margins of vitiligo lesions.[48] In addition, perilesional melanocytes consistently express MHC class II antigens and ICAM-1 which have a role in the activation of helper T lymphocytes.[49] Specific cellular immune responses to melanocyte antigens have recently been demonstrated: the presence of circulating melan-A-specific and tyrosinase-specific cytotoxic T lymphocytes has been shown in a significant number of vitiligo patients when compared to control subjects.[50] The T cells expressed high levels of the skin-homing receptor cutaneous lymphocyte-associated antigen and their frequency correlated with the extent of depigmentation.[50] These findings are consistent with a role for skin-homing melanocyte-specific T lymphocytes in the pathogenesis of vitiligo, and the study was the first to identify specific T cell reactivities to melanocytes in this depigmenting disease.

With regard to genetic factors that may predispose to the development of vitiligo, there is a family history in over 20% of patients.[51] However, although the pattern of relationship between relative risk and degree of kinship indicates the involvement of genetic factors, this is not consistent with a simple mendelian mode of inheritance and it is therefore likely that the disorder is due to the action of genes at multiple loci.[51] Several studies have reported associations of vitiligo with certain HLA specificities, but there is no consistent association with any of the MHC class I or class II alleles.[52]

The results of a recent case-control study revealed no association between the 106 base pair allele of the CTLA-4 microsatellite $(AT)_n$ repeat polymorphism and vitiligo in patients without an autoimmune disorder (Table 5).[53] In contrast, the frequency of the 106 base pair allele was significantly increased in the group of vitiligo patients with an autoimmune disease (Table 5). The fact that 81% of patients in this subgroup had other disorders previously associated with the 106 base pair allele (e.g., Graves' disease, autoimmune hypothyroidism and autoimmune Addison's disease), probably accounts for the apparent association of vitiligo and this particular CTLA-4 polymorphic allele. However, the relative risk of 3.9, conferred by the 106 base pair allele in this patient group, was greater than that found for patients with only Graves' disease,[13] autoimmune hypothyroidism[13] and autoimmune Addison's disease[54] with relative risk values of 2.1, 2.2 and 2.2, respectively. This suggests that autoimmune endocrinopathy patients with the 106 base pair allele may have

Table 5. CTLA-4 106 base pair $(AT)_n$ repeat allele frequency in vitiligo patients[1,2]

	Controls (n = 173)	Vitiligo[3] (n = 74)	Vitiligo[4] (n = 53)	Vitiligo[5] (n = 21)
Number of alleles investigated	346	148	106	42
Number of 106 base pair alleles	47	33	17	16
Frequency of the 106 base pair allele	13.6%	22.3%	16.0%	38.1%
P value[6]	-	0.017	0.633	0.0001
Odds ratio[6]		1.83	1.22	3.92

[1] n, number of individuals.
[2] Data from Kemp et al [53].
[3] All patients with vitiligo.
[4] Patients with vitiligo and no other autoimmune disease.
[5] Patients with vitiligo and at least one other autoimmune disease.
[6] Compared with control group.

a greater susceptibility to the development of vitiligo. Examination of the clinical subclass of vitiligo (e.g., symmetrical, segmental) revealed no association of vitiligo type and the 106 base pair allele.[53] In addition, no association was found between the age of onset of the disease and the 106 base pair genotype.[53] The results reported in this initial study have been confirmed with respect to the microsatellite marker using a larger cohort of vitiligo patients, but an association between the A/G polymorphism and vitiligo was not evident.[55]

Conclusion

There now seems little doubt that polymorphisms of the CTLA-4 gene are associated with both Graves' disease and autoimmune hypothyroidism, and are second only to HLA polymorphisms in the strength of their assocations. The evidence is less clear for vitiligo which frequently occurs in patients with autoimmune thyroid disease and other endocrinopathies. More work is required in this area, but it seems likely that vitiligo with a strong autoimmune component etiologically does have a positive association with CTLA-4 polymorphisms. As these polymorphisms have also been associated with type 1 diabetes mellitus and Addison's disease, there seems to be a similar relationship to polyendocrine autoimmunity as exists with the HLA-DR3 specificity which is found in association with a broad range of conditions.

What is less clear at present is the reason for the association; the essential question is whether the polymorphisms in the CTLA-4 gene confer a property on the expressed protein which predispose to autoimmunity, or whether the associations reflect nothing more than linkage disequilibrium with CD28 or with a close and unidentified gene. The last possibility is counter-intuitive, and a direct relationship with CTLA-4 itself is the more likely explanation by recent in vitro studies: peripheral blood T cells from individuals with the position 49 G/G alleles showed reduced augmentation of their proliferative response to allogenic B cells in the presence of blocking antibodies to CTLA-4, compared with those from A/A individuals [56]. Thus, the CTLA-4 encoded by G/G alleles appears to cause less suppression of T cells, and, in vivo, this may lead to the expansion of autoreactive T cells. In this way, an appropriate CTLA-4 polymorphism would act as a non-specific enhancer of the autoimmune response.

References

1. Brix TH, Kyvik KO, Hegedus L. What is the evidence of genetic factors in the etiology of Graves' disease? A brief review. Thyroid 1998; 8:727-732.
2. Weetman AP. Graves' disease. New Engl J Med 2000; 343:1236-1248.
3. Yanagawa T, Hidaka Y, Guimaraes V et al. CTLA-4 gene polymorphism associated with Graves' disease in a Caucasian population. J Clin Endocrinol Metab 1995; 80:41-45.
4. Vaidya B, Imrie H, Perros P et al. The cytotoxic T lymphocyte antigen-4 is a major Graves' disease locus. Hum Mol Genet 1999; 8:1195-1199.
5. Barbesino G, Tomer Y, Concepcion E et al. The International Consortium for the Genetics of Autoimmune Thyroid Disease. Linkage analysis of candidate genes in autoimmune thyroid disease: 1. Selected immunoregulatory genes. J Clin Endocrinol Metab 1998; 83:1580-1584.
6. Heward JM, Allahabadia A, Armitage M et al. The development of Graves' disease and the CTLA-4 gene on chromosome 2q33. J Clin Endocrinol Metab 1999; 84:2398-2401.
7. Djilali-Saiah I, Larger E, Harfouch-Hammoud E et al. No major role for the CTLA-4 gene in the association of autoimmune thyroid disease with IDDM. Diabetes 1998; 47:125-127.
8. Bahn RS, Heufelder AE. Pathogenesis of Graves' ophthalmopathy. N Engl J Med 1993; 329:1468-1475.
9. Perros P, Crombie AL, Matthews JNS et al. Age and gender influence the severity of thyroid-associated ophthalmopathy. Clin Endocrinol 1993; 38:367-372.
10. Shine B, Fells P, Edwards OM et al. Association of Graves' ophthalmopathy and smoking. Lancet 1990; 335:1261-1263.
11. Vaidya B, Imrie H, Perros P et al. Cytotoxic T lymphocyte antigen-4 (CTLA-4) gene polymorphism confers susceptibility to thyroid associated ophthalmopathy. Lancet 1999; 354:743-744.
12. Buzzetti R, Nistico L, Signore A, Cascino I. CTLA-4 and HLA gene susceptibility to thyroid-associated orbitopathy. Lancet 1999; 354:1824.
13. Kotsa K, Watson PF, Weetman AP. A CTLA-4 gene polymorphism is associated with both Graves' disease and autoimmune hypothyroidism. Clin Endocrinol 1997; 46:551-554.
14. Villanueva R, Inzerillo AM, Tomer Y et al. Limited susceptibility to severe Graves' ophthalmopathy: no role for CTLA-4 but evidence for an environmental etiology. Thyroid 2000; 10:791-798.
15. Tunbridge WMG, Evered DC, Hall R et al. The spectrum of thyroid disease in the community: the Wicham survey. Clin Endocrinol 1977; 7:481-492.
16. Kaufman KD, Filetti S, Seto P et al. Recombinant human thyroid peroxidase generated in eukaryotic cells: a source of specific antigen for the immunological assay of antimicrosomal antibodies in the sera of patients with autoimmune thyroid disease. J Clin Endocrinol Metab 1990; 70:724-728.
17. Tandon N, Zhang L, Weetman AP. HLA associations in Hashimoto's thyroiditis. Clin Endocrinol 1991; 34:383-386.
18. Donner H, Braun J, Seidl C et al. Codon 17 polymorphism of the cytotoxic T lymphocyte antigen 4 gene in Hashimoto's thyroiditis and Addison's disease. J Clin Endocrinol Metab 1997; 82:4131-4132.
19. Awata T, Kurihara S, Iitaka M et al. Association of CTLA-4 gene A-G polymorphism (IDDM12 locus) with acute-onset and insulin-depleted IDDM as well as autoimmune thyroid disease (Graves' disease and Hashimoto's thyroiditis) in the Japanese population. Diabetes 1998; 47:128-129.
20. Akamizu T, Sale MM, Rich SS et al. Association of autoimmune thyroid disease with microsatellite markers for the thyrothropin receptor gene and CTLA-4 in Japanese patients. Thyroid 2000; 10:851-858.
21. Park YJ, Chung HK, Park DJ et al. Polymorphism in the promoter and exon 1 of the cytotoxic T lymphocyte antigen 4 gene associated with autoimune thyroid disease in Koreans. Thyroid 2000; 10:453-459.
22. Hewerd JM, Allahabadia A, Carr-Smith J et al. No evidence for allelic association of a human CTLA-4 promoter polymorphism with autoimmune thyroid disease in either population-based case-control or family-based studies. Clin Endocrinol 1998; 49:331-334.

23. Braun J, Donner H, Siegmund T et al. CTLA-4 promoter variants in patients with Graves' disease and Hashimoto's thyroiditis. Tissue Antigens 1998; 51:563-566.
24. Lazarus JH, Othman S. Thyroid disease in relation to pregnancy. Clin Endocrinol 1991; 34:91-98.
25. Othman S, Phillips DIW, Parkes AB et al. A long-term follow-up of PPT. Clin Endocrinol 1990; 32:559-564.
26. Feldt-Rasmussen U, Hoier-Madsen M, Rasmussen NG et al. Anti-thyroid peroxidase antibodies during pregnancy and postpartum. Relation to PPT. Autoimmunity 1990; 6:21-214.
27. Jansson R, Totterman TH, Sallstrom J et al. Intrathyroidal and circulating lymphocyte subsets in different stages of autoimmune postpartum thyroiditis. J Clin Endocrinol Metab 1984; 58:942-946.
28. Rajatanavin R, Chailurkit L, Tirarungsikul K et al. Postpartum thyroid dysfunction in Bangkok: a geographical variation in prevalence. Acta Endocrinol 1990; 122:283-287.
29. Parkes AB, Darke C, Othman S et al. Major histocompatibility complex class II and complement polymorphisms in postpartum thyroiditis. Eur J Endocrinol 1996; 134:449-453.
30. Waterman EA, Watson PF, Lazarus JH et al. A study of the association between a polymorphism in the CTLA-4 gene and postpartum thyroiditis. Clin Endocrinol 1998; 49:251-255.
31. Ochi Y, DeGroot LJ. Vitiligo in Graves' disease. Ann Intern Med 1969; 71:935-940.
32. Macaron C, Winter RJ, Traisman HS et al. Vitiligo and juvenile diabetes mellitus. Arch Dermatol 1977; 113:1515-1517.
33. Naughton GK, Eisinger M, Bystryn J-C. Antibodies to normal human melanocytes in vitiligo. J Exp Med 1983;158:246-251.
34. Bystryn J-C, Naughton GK. The significance of vitiligo antibodies. J Dermatol 1985; 12:1-9.
35. Naughton GK, Reggiardo MD, Bystryn J-C. Correlation between vitiligo antibodies and extent of depigmentation in vitiligo. J Am Acad Dermatol 1986; 15:978-981.
36. Harning R, Cui J, Bystryn J-C. Relation between the incidence and level of pigment cell antibodies and disease activity in vitiligo. J Invest Dermatol 1991; 97:1078-1080.
37. Norris DA, Kissinger RM, Naughton GK et al. Evidence for immunologic mechanisms in human vitiligo: patients' sera induce damage to human melanocyes in vitro by complement-mediated damage and antibody-dependent cellular cytotoxicity. J Invest Dermatol 1988; 90:783-789.
38. Gilhar A, Zelickson B, Ulman Y et al. In vivo destruction of melanocytes by the IgG fraction of serum from patients with vitiligo. J Invest Dermatol 1995; 105:683-686.
39. Cui J, Harning R, Henn M et al. Identification of pigment cell antigens defined by vitiligo antibodies. J Invest Dermatol 1992; 98:162-165.
40. Cui J, Arita Y, Bystryn J-C. Characterisation of vitiligo antigens. Pigment Cell Res 1995; 8:53-59.
41. Song YH, Connor E, Li Y et al. The role of tyrosinase in autoimmune vitiligo. Lancet 1994; 344:1049-1052.
42. Baharav E, Merimski O, Shoenfeld Y et al. Tyrosinase as an autoantigen in patients with vitiligo. Clin Exp Immunol 1996; 105:84-88.
43. Kemp EH, Gawkrodger DJ, MacNeil S et al. Detection of tyrosinase autoantibodies in vitiligo patients using ^{35}S-labelled recombinant human tyrosinase in a radioimmunoassay. J Invest Dermatol 1997; 109:69-73.
44. Kemp EH, Waterman EA, Gawkrodger DJ et al. Autoantibodies to tyrosinase-related protein-1 detected in the sera of vitiligo patients using a quantitative radiobinding assay. Br J Dermatol 1998; 139:798-805.
45. Okamoto T, Irie RF, Fujii S et al. Anti-tyrosinase-related protein-2 immune response in vitiligo and melanoma patients receiving active-specific immunotherapy. J Invest Dermatol 1998; 111:1034-1039.
46. Kemp EH, Gawkrodger DJ, Watson PF et al. Immunoprecipitation of melanogenic enzyme autoantigens with vitiligo sera: evidence for cross-reactive autoantibodies to tyrosinase and tyrosinase-related protein-2 (TRP-2). Clin Exp Immunol 1997; 109:495-500.
47. Kemp EH, Gawkrodger DJ, Watson PF et al. Autoantibodies to human melanocyte-specific protein Pmel17 in the sera of vitiligo patients: a sensitive and quantitative radioimmunoassay (RIA). Clin Exp Immunol 1998; 114:333-338.
48. LePoole IC, van den Wijngaard RMJGJ, Westerhof W et al. Presence of T cells and macrophages in inflammatory vitiligo skin parallels melanocyte disappearance. Am J Pathol 1996; 148:1219-28.
49. Al Badri AMT, Foulis AK, Todd PM et al. Abnormal expression of MHC class II and ICAM-1 by melanocytes in vitiligo. J Pathol 1993; 169:203-206.
50. Ogg GS, Dunbar PR, Romero P et al. High frequency of skin-homing melanocyte-specific cytotoxic T lymphocytes in autoimmune vitiligo. J Exp Med 1998; 188:1203-1208.

51. Majumder PP, Nordlund JJ, Nath SK. Pattern of familial aggregation of vitiligo. Arch Dermatol 1993; 129:994-998.
52. Schallreuter KU, Levenig C, Kuhnl P. Histocompatability antigens in vitiligo: Hamburg Study on 102 patients from Northern Germany. Dermatology 1993; 187:186-192.
53. Kemp EH, Ajjan RA, Waterman EA et al. Analysis of a microsatellite polymorphism of the cytotoxic T lymphocyte antigen-4 gene in patients with vitiligo. Br J Dermatol 1999; 140:73-78.
54. Kemp EH, Ajjan RA, Husebye ES et al. A cytotoxic T lymphocyte antigen-4 (CTLA-4) gene polymorphism is associated with autoimmune Addison's disease in English patients. Clin Endocrinol 1998; 49:609-613.
55. Kristensen DL, Bradley CA, Hao M et al. Case-control association study of candidate genes for vitiligo susceptibility: tyrosine metabolism and immune response genes. Clin Chem 2000; 46 (Part 2 Supplement):811 (Abstract).
56. Kouki T, Sawai Y, Gardine CA et al. CTLA-4 gene polymorphism at position 49 in exon 1 reduces the inhibitory function of CTLA-4 and contributes to the pathogenesis of Graves' disease. J Immunol 2000; 165:6606-6611.
57. Donner H, Rau H, Walfish PG et al. CTLA-4 alanine-17 confers genetic susceptibility to Graves' disease and to type 1 diabetes mellitus. J Clin Endocrinol Metab 1997; 82:143-146.
58. Yanagawa T, Taniyama M, Enomoto S et al. CTLA-4 gene polymorphism confers susceptibility to Graves' disease in Japanese. Thyroid 1997; 7:843846.

Chapter 6

CTLA-4 in Myasthenia Gravis

Ann Kari Lefvert

Introduction

Myasthenia gravis (MG) is commonly regarded as the prototype for an organ specific antibody-mediated autoimmune disease. The disease is characterized by an immune response against the nicotinic acetylcholine receptor on the neuromuscular junction. The symptoms, weakness and increased fatigability, are considered to be caused by direct blockade and a reduction in the number of functional receptors at the neuromuscular junction by the autoantibodies.[1]

The poor correlation between the autoantibody concentration and the disease severity challenges the concept of a simple cause and effect relationship between these antibodies and the disease. Moreover, acetylcholine receptor antibodies are found in several conditions not accompanied by neuromuscular symptoms, including some thymomas,[2] healthy first-degree relatives[3] and in the healthy twin in a monozygotic pair of twins discordant for MG,[4] in monoclonal gammopathies[5] and in primary biliary cirrhosis.[6]

Recent reports suggest that proinflammatory cytokines, such as IL-1 and tumor necrosis factor TNF, may play a more direct role in the development of the disease and its symptoms. Transgenic mice that express IFN-γ at the neuromuscular junction, develop a myasthenia-like disease.[7] Patients with MG, especially those with no HLA-B8 association, have a rather strong genetic association to high-secretory genotype A2 of IL-1β,[8] and to the high-secretory genotype A2 of TNF-α.[9] Animal experiments have supported the important role of IL-1β. Mice deficient of IL-1β have a much-reduced incidence and disease-severity of experimental myasthenia gravis induced by acetylcholine receptor. In these mice, both T- and B- cell responses to the receptor are reduced.[10]

Further evidence supporting the importance of a proinflammatory mechanism in MG is the inflammatory reaction at the endplate and mononuclear cell infiltration in the skeletal muscles in some patients, especially those with thymic tumors.[11] Treatment of patients with anti-CD4+ antibodies resulted in long-lasting remission and abolished T cell reactivity to the autoantigen, without any decrease of concentration of the autoantibodies.[12]

In all, these results suggest that mechanisms facilitating a pro-inflammatory reaction might be of importance in MG. CTLA-4 is an essential component of the immune system and serves as a negative regulator for T cell activation. Animals deficient for CTLA-4 show a massive T-cell lymphoproliferative disorder with increased numbers of activated T-cells and elevated basal levels of serum immunoglobulins, resulting in autoimmune-like tissue destruction.[13,14] Defect expression/function of CTLA-4 should thus result in abnormal T-cell activation and an exaggerated inflammatory/immune response.

CTLA-4 in Autoimmune Disease, edited by Flemming Pociot. ©2004 Eurekah.com.

We have analyzed the distribution of *Ctla-4* polymorphisms in Swedish MG patients compared to healthy individuals, the relation of these polymorphisms to thymic histopathology and to other genetic markers and the functional consequences of certain genotypes.

Distribution of *Ctla-4* (AT)n Polymorphism in Patients with MG

The gene for CTLA-4 (CTLA4) is located on the human chromosome 2q33 and is closely linked to the CD28 gene. The CTLA4 includes an $(AT)_n$ microsatellite within the 3′- untranslated region of exon 3.

In the Swedish Caucasian population, we observed totally 16 alleles, with sizes ranging from 86 to 128 bp.[15] There was no difference in the allelic distribution between healthy controls and MG patients as a whole. The most common genotype in patients was 86/86 bp, present in 56.1 %, followed by 104/104 bp, present in 14.6%. The frequencies of these genotypes in healthy individuals were the same, 52% and 18%, respectively. Some of the rare genotypes were presented in either inpatients or in healthy controls. These differences were significant ($p < 0.0001$).[15]

Thus, the shortest and the most common CTLA4 allele in the Swedish Caucasian population was 86 bp. This is in accordance with the results from previous studies in which the most common allele is also the shortest allele.[16]

Ctla-4 (AT)n Is Associated with Thymoma

Table 1 presents the allelic frequencies in patients with normal thymic histology, thymoma and thymic hyperplasia. The frequency of the shortest allele, 86 bp, in patients with thymoma was decreased compared to patients with hyperplasia and normal thymic histology. The prevalence of allele 104 in patients with thymoma had a tendency to be increased ($p = 0.088$) as compared to patients with hyperplasia. There was no difference between unthymectomized patients and healthy individuals.

A few reports have described associations of MHC genes in patients with thymoma to be different from that in patients with hyperplasia or normal thymic histology.[17,18] Patients with thymomas constitute a distinct clinical entity in which the disease usually is severe, starts in middle age and is as common in men as in women. These patients frequently have autoantibodies against muscle proteins and mononuclear cell infiltrations in the skeletal muscles.[19] In this regard, it is of special interest to note that mice deficient for the CTLA4 have a massive lymphocyte infiltration in skeletal muscles.

CD28 and CTLA-4 and their ligands are expressed in the thymus and may have a role in the induction of T-cell tolerance.[20,21] Mice deficient for CTLA4 exhibit abnormal composition of thymocytes, consisting of higher percentages of single positive cells ($CD4^+$ $CD8^-$ or $CD4^-$ $CD8^+$) and lower percentages of double positive cells ($CD4^+$ $CD8^+$).[14,22] Insufficient expression of CTLA-4 in thymus due to genetic variation in the CTLA4 region may decrease the avidity in the interactions between immature thymocytes and antigen-presenting cells, resulting in maturation of self-reactive T cells.[23] Higher percentages of longer alleles of CTLA4 in persons who develop MG and thymoma could affect thymic selection, thus triggering an autoimmune response. Recent studies indeed suggest that abnormal thymic selection exists in the thymus from MG patients.[24,25] This association between CTLA4 and MG with thymoma further confirms the hypothesis that the pathogenesis of MG in patients with thymoma appears to be different from that in other subgroups.

Table 1. CTLA-4 (AT)n in healthy individuals and in myasthenia gravis patients with different thymic histopathology. The numbers are given as percentages.

Allele	MG	Healthy	Thymic Histopathology			
			Thymoma	Hyperplasia	Normal	No tx
86	61.5	58.5	**42.5***	69	73.1	58.5
90	0.8	NP	NP	NP	NP	2.1
94	0.4	0.5	NP	1.0	NP	NP
102	2.7	1.0	5.0	1.0	NP	4.3
104	25.0	26.5	**37.5**	22.0	26.9	22.3
106	1.5	2.5	2.5	NP	NP	3.2
108	1.9	1.5	2.5	2.0	NP	2.1
110	1.9	1.5	5.0	1.0	NP	2.1
112	1.1	2.0	NP	NP	NP	2.1
116	0.4	2.5	2.5	NP	NP	NP
118	0.8	0.5	2.5	1.0	NP	NP
120	0.4	0.5	NP	1.0	NP	NP
122	0.4	NP	NP	1.0	NP	NP
124	NP	0.5	NP	NP	NP	NP
126	0.4	1.5	NP	1.0	NP	NP
128	0.8	0.5	NP	NP	NP	2.1

*$p < 0.05$, OR = 0.272, 95% confidence intervall=0.093-0.794, as compared to patients with normal thymic histopathology

*$p < 0.01$, OR = 0.332, 95% confidence intervall=0.156-0.708, as compared to patients with thymic hyperplasia.

Different Distribution of *Ctla-4* (AT)n in MG Patients Stratified by Genetic Markers on Chromosomes 2 and 6

To uncover the possible contribution of CTLA4 in different subgroups of MG patients, the patients were divided into subgroups according to TNF-α -308 *NcoI* RFLP A2 and IL-1β *TaqI* A2. Table 2 shows the frequencies of the CTLA4 alleles in patients stratified by those two genetic markers. The percentage of allele 104 was increased in patients with the IL-1β $A2^{+}$ phenotype compared to the patients negative for this allele. The prevalence of allele 86 showed a tendency to be decreased in patients with the IL-1β $A2^{+}$ patients ($p = 0.0822$). The percentage of allele 104 was somewhat lower in patients positive for TNFα -308 A2, while higher in patients negative for TNFα -308 A2, in which the majority of patients are positive for IL-1β *TaqI* A2.

We have earlier shown that IL-1β *TaqI* RFLP A2 is associated with myasthenia gravis,[8] especially in patients negative for HLA-B8. IL-1β *TaqI* RFLP A2 is in strong linkage with IL-1β *NcoI* RFLP A2, which is associated with patients with normal thymus and without serum acetylcholine receptor antibodies. The percentage of CTLA4 allele 104 is higher in patients positive for IL-1β *TaqI* RFLP A2, an IL-1β "high-secretor" phenotype. The distance between IL-1β *TaqI* RFLP, $2q^{13}$, and CTLA4, $2q^{33}$ is more than 10 cM,[26] and the association of CTLA4 should not be due to variation at the IL-1β/IL-1α locus. Thus, variations in CTLA4 may confer susceptibility to the disease without the contribution from IL-1 genes. Alternatively, CTLA4 and IL-1 genes may act synergistically. The pathogenic mechanism of the disease associated with IL-1 gene, but not MHC gene

Table 2. CTLA-4 polymorphisms in myasthenia gravis patients stratified by genetic markers of IL-1β* TacI *and TNF-α* NcoI *RFLP A2. Numbers are expressed as percentages.

Allele	IL-1β RFLP A2+	IL-1β RFLP A2-	TNF-α RFLP A2+	TNF-α RFLP A2-
86	59.8	72.4	74.5	56.0
90	NP	NP	NP	NP
94	1.0	NP	NP	1.0
102	NP	3.0	1.0	2.0
104	**30.4***	14.3	17.3	28.0
106	2.9	NP	2.2	1.0
108	1.0	3.1	1.0	3.0
110	2.0	2.0	1.0	3.0
112	1.0	NP	1.0	NP
116	NP	1.0	NP	1.0
118	NP	2.0	1.0	1.0
120	NP	1.0	NP	1.0
122	NP	1.0	NP	1.0
126	NP	1.0	1.0	NP
128	2.0	NP	NP	2.0

RFLP denotes restriction fragment length polymorphism.
*: $p<0.05$, OR = 2.619, 95% confidence intervall = 1.293-5.307, as compared to MG patients not having IL-1β *TacI* RFLP allel A2.

(e.g., TNF-α -308 A2) associated mechanism, can be potentiated by the presence of CTLA4 disease associated alleles. In this regard, it is of interest to note that mice expressing either TNFα or CD80 in the pancreatic B cells do not develop diabetes, but littermates coexpressing TNFα and CD80 do develop a severe disease.[27] The fact that CTLA4 -/- mice have lymphocyte infiltration in the skeletal muscle[22] provides further evidence that patients who have longer CTLA4 genes may have an additional pathogenic mechanism in addition to antibody mediated immune response against the acetylcholine receptor, such as destruction of neuromuscular junction directly by macrophages, T lymphocytes and/or their products (e.g., IL-1). Alternatively, variations in CTLA4 microsatellite can serve as markers for other sinful mutation(s) in the coding or regulatory regions of CTLA4, CD28 gene, or unknown genes in linkage disequilibrium with CTLA4.

Ctla-4 (AT)n Is Associated to ADCC

A possible additional mechanism for pathogenic action of the acetylcholine receptor antibodies might be antibody-dependent cell-mediated cytotoxicity (ADCC). Using a cell line expressing nicotinic acetylcholine receptor as target cells, we could demonstrate increased ADCC mediated by sera from MG patients compared to sera from healthy individuals ($p<0.0001$).[28] Sera with autoantibodies induced a higher cytotoxicity than sera from patients without. Sera from MG patients with thymomas induced a higher cytotoxic effect than sera from other patients. Sera from thymoma patients who had extended dinucleotide repeats, (AT)n repeats, in the CTLA-4 gene mediated especially high cytotoxicity ($p<0.05$). Sera from thymoma patients with both (AT)n in the CTLA-4 gene longer than 86 bp had significantly higher cytotoxic effects when compared with patients homozygous for 86/86 or heterozygous for 86 and an allele >86 ($p<0.02$). Antibody-dependent cell-mediated cytotoxicity mediated by acetylcholine receptor antibodies may thus be another possible pathogenic mechanism that could operate in MG patients, especially in patients with thymomas.

Hence, our results suggested that ADCC via acetylcholine receptor antibodies might be another possible way of causing a loss of the receptor at the endplate, especially in patients with thymomas. Of interest is our result that sera from patients with longer (AT)n repeats in the CTLA-4 gene had a significantly higher cytotoxicity when compared with thymoma patients with shorter (AT)n repeats. Thymoma patients generally have higher levels of acetylcholine receptor autoantibodies.[29] They have a high prevalence of infiltration of mononuclear cells in skeletal muscles and often other autoantibodies against skeletal muscle components.[19,30] Patients with longer (AT)n repeats in the gene have an increased T cell activation by the CD28 pathway compared to patients having CTLA-4 allele with a length of 86 bp, and may thus display a more active cell mediated immunity, including ADCC.[31] The Th1 type cytokines TNF-α and IL-1 are cytotoxic and may be involved directly in the lytic machinery,[32] IL-8 can trigger superoxide anion release[33] and IFN-γ, IL-4, IL-10 and IL-13 modulate the expression of Fc receptors.[34] These mechanisms might be of importance for the increase in ADCC in patients having long (AT)n repeats in the CTLA-4 gene and thus an increased T cell activation and secretion of cytokines.

Functional Correlates to the *Ctla-4* (AT)n

In order to determine the functional consequences of different (AT)n genotypes, we performed a study of T-cell activation in patients with myasthenia gravis not treated with immunosuppressive drugs.[31] Three parameters of T-cell activation are presented in Table 3. One parameter of T-cell activation is the concentration of IL-2 sRα in serum. This increased in parallel to the length of (AT)n in CTLA4 . Patients with CTLA4 >86/>86 had higher levels of IL-2 sRα than patients with CTLA4 86/86, while patients with CTLA4 86/>86 had an intermediate level of serum IL-2 sRα. Another parameter for T cell activation is the activity of telomerase. Patients with (AT)n >86/>86 bp had higher levels of telomerase activity in their PBMC than patients with CTLA4 86/86 bp. Stimulation by the acetylcholine receptor in vitro slightly increased the activity of telomerase expressed by PBMC When the CTLA-4 activity in PBMC was blocked using anti-CTLA-4 antibodies, the autoantigen-induced stimulation was significantly augmented. There was no difference in the spontaneous lymphocyte proliferation among patients with different CTLA4 genotypes. However, addition of anti-CD28 antibodies to anti-CD3-containing cultures resulted in significantly higher proliferation than in cultures with anti-CD3 alone. Of particular interest was that cells from patients with CTLA4 >86/>86 had a significantly higher increase of T cell proliferation with the addition of anti-CD28 to the anti-CD3 culture system than cells from patients with CTLA4 86/86. Cells from patients with CTLA4 86/>86 showed an intermediate increase (Table 3).

Conclusions

In myasthenia gravis, there is a genetic correlation to polymorphisms of the proinflammatory cytokines IL-1β and TNF-α, as well as a correlation of certain genotypes of IL-10.[35] There is, however, no relation to polymorphisms of the genes for IL-4 and IL-6.[36,37] This suggests that proinflammatory mechanisms might be of importance

Table 3. Signs of T cell stimulation in individuals bearing (AT)n 86/86 bp, 86/>86 bp and >86/<86 bp

Genotype	Median	95% Confidence Intervall
A		
86/86*	480	380-750
86/>86	700	470-800
>86/>86	750	500-1300
B		
86/86**	0.16	0.04-0.30
>86/<86	0.23	0.12-0.67
C		
86/86#	100	0-600
86/>86	500	200-900
>86/>86	1500	500-4500

A: Soluble IL-2R in serum expressed in pg/ml; B: Telomerase activity in PBMC expressed in optical density units; C: ^{3}H-Thymidine incorporation in PBMC cultured together with anti-CD3 + anti-CD28, expressed in CPM; *$p<0.001$ compared to persons having >86/>86; **$p<0.05$ compared to persons having >86/>86; #$p<0.005$ compared to persons having >86/>86.

for MG. This is further suggested by results from animal experiments.[10] The function of CTLA-4 is to oppose CD28, and thus to prevent an over-activation of T cells. The longer (AT)n in the gene is clearly related to increased T cell activation in MG.[31] Generally, AT rich regions in the 3′untranslated regions have influence on mRNA stability, especially in cytokine genes.[38] When a nucleotide AT sequence is inserted into the 3′-untranslated region of rabbit β-globin gene, stable wild-type mRNA becomes highly unstable. No stable transcript with long AU sequence expressed from coding or noncoding regions of genes has been found. The lethal phenotype of animal homozygous for null mutation in CTLA4 strongly supports a critical role for CTLA-4 in down regulating T cell activation and maintaining immunological homeostasis.

It is reasonable to speculate that mRNA from the shorter alleles of CTLA4 might be more stable than that from longer alleles.[16] This has also now recently been shown in our laboratory.[39] The shorter alleles could thus be regarded as disease-protective, the longer alleles as disease-associated alleles. Thus, decreased percentage of disease-protective and increased prevalence of disease-associated alleles in patients with thymomas might lead to an unstability of CTLA4 mRNA, a defect of CTLA4 expression and thus an imbalance between the positive (CD28) and the negative (CTLA-4) co-stimulation. The insufficiently opposed CD28 signal pathway triggers T- and B- cell activation and autoantibody secretion in vivo, resulting in propagation of the disease.

References

1. Drachman DB. Myasthenia gravis. New Eng J Med 1994; 330:1797-1810.
2. Aarli JA, Lefvert AK, Tonder O. Thymoma-specific antibodies in sera from patients with myasthenia gravis demonstrated by indirect haemagglutination. J Neuroimm 1981; 1:421-427
3. Lefvert AK, Pirskanen R, Svanborg E. Anti-idiotypic antibodies, acetylcholine receptor antibodies and disturbed neuromuscular function in healthy relatives to patients with myasthenia gravis. J Neuroimm 1985; 9:41-53

4. Lefvert AK, Pirskanen R, Eng H et al. B cell and autoantibody repertoire in a pair of monozygotic twins discordant for myasthenia gravis. Clin Immunol Immunopathol 1989; 53:161-170
5. Eng H, Lefvert AK, Mellstedt H et al. Human monoclonal immunoglobulins that bind the human acetylcholine receptor. Eur J Immunol 1987; 17:1867-1869
6. Sundewall AC, Lefvert AK, Olsson R. Anti- acetylcholine receptor antibodies in primary biliary cirrhosis. Acta Med Scand 1985; 217:519-525
7. Danling G, Wogensen L, Calcutt NA et al. Myasthenia gravis-like syndrome induced by expression of interferon γ in the neuromuscular junction. J Exp Med 1995; 181:547-557.
8. Huang D, Pirskanen P, Hjelmström P et al. Polymorphisms in IL-1b and IL-1 receptor antagonist genes are associated with myasthenia gravis. J Neuroimm 1998; 81:76-81.
9. Huang D, Pirskanen R, Matell G et al. Tumour necrosis factor-α polymorphism and secretion in myasthenia gravis. J Neuroimm 1999;94:165-171.
10. Huang D, Shi FD, Giscombe R et al. Disruption of the Il-1β gene diminishes acetylcholine receptor-induced immune responses in a murine model of myasthenia gravis. Eur J Immunol 2001; 31:225-232.
11. Maselli RA, Richman DP, Wollman RL. Inflammation at the neuromuscular junction in myasthenia gravis. Neurology 1991; 41:1497-1504.
12. Åhlberg R, Yi Q, Pirskanen R et al. Treatment of myasthenia gravis with anti-CD4 antibody: Improvement correlates to decreased T-cell autoreactivity. Neurology 1994; 44:1732-1737.
13. Tivol EA, Borriello F, Schweitzer AN et al. Loss of CTLA-4 leads to massive lymphoproliferation and fatal multiorgan tissue destruction, revealing a critical negative regulatory role of CTLA-4. Immunity 1995; 3:541-547.
14. Waterhouse P, Penninger JM, Timms E et al. Lymphoproliferative disorders with early lethality in mice deficient in *Ctla-4*. Science 1995; 270:985-988.
15. Huang D, Liu L, Norén K et al. Genetic association of CTLA-4 to myasthenia gravis with thymoma. J Neuroimmunology 1998; 88:192-198.
16. Yanagawa T, Hidaka Y, Guimaraes V et al. CTLA-4 gene polymorphism associated with Grave's disease in a caucasian population. J Clin Endocrin 1995; 80:41-45.
17. Carlsson B, Wallin J, Pirskanen R et al. Different HLA DR-DQ associations in subgroups of idiopathic myasthenia gravis. Immunogenetics 1990; 31:285-290.
18. Spurkland A, Gilhus NE, Ronningen KS et al. Myasthenia gravis patients with thymus hyperplasia and myasthenia gravis patients with thymoma display different HLA associations. Tissue Antigens 1991; 37:90-93.
19. Aarli JA, Stefansson K, Marton LSG et al. Patients with myasthenia gravis and thymoma have in their sera IgG autoantibodies against titin. Clin Exp Immunol 1990; 82:284-288.
20. Degermann S, Surh CD, Glimcher LH et al. B7 expression on thymic medullary epithelium correlates with epithelium-mediated deletion of $V\beta5^+$ thymocytes. J Immunol 1994; 152:3254-3263.
21. Cibotti R, Punt JA, Dash KS et al. Surface molecules that drive T cell development in vitro in the absence of thymic epithelium and in the absence of lineage-specific signals. Immunity 1997; 6:245-255.
22. Tivol EA, Boyd SD, Mc Keon S et al. CTLA-4 Ig prevents lymphoproliferation and fatal multiorgan tissue destruction in CTLA-4-deficient mice. J Immunol 1997; 158:5091-5094.
23. Liu Y. Is CTLA-4 a negative regulator for T-cell activation? Immunol Today 1997; 18:569-572.
24. Truffault F, Cohen-Kaminsky S, Khalil I et al. Altered intrathymic T-cell repertoire in human myasthenia gravis. Ann Neurol 1997; 41:731-741.
25. Moulian N, Bidault J, Truffault F et al. Thymocyte Fas expression is dysregulated in myasthenia gravis patients with anti-acetylcholine receptor antibody. Blood 1997; 89:3287-3295.
26. Lafage M, Maroc N, Dubreuil P et al. The human interleukin-1α gene is located on the long arm of chromosome 2 at band q13. Blood 1989; 73:104-107.
27. Guerder S, Picarella DE, Linsley S et al. Costimulator B7-1 confers antigen-presenting-cell function to parenchymal tissue and in conjunction with tumor necrosis factor a leads to autoimmunity in transgenic mice. Proc Natl Acad Sci USA 1994; 91:5138-5142.
28. Xu BY, Pirskanen R, Lefvert AK. Antibody- dependtdent cytotoxicity: an additional mechanism that might operate in human myasthenia gravis J Neuroimm 1999; 99:183-188.
29. Lefvert AK, Bergström K, Matell G et al. Determination of acetylcholine receptor antibody in myasthenia gravis: clinical usefulness and pathogenetic implications. J Neurol Neurosurg Psychiat 1978; 41:394-403.

30. Mygland Å, Tysnes O-B, Matre R et al. Anti-cardiac ryanodine receptor antibodies in thymoma-associated myasthenia gravis. Autoimm 1994; 17:327-331.
32. Pullybland AM, Guillow PH, Monson JRT. Interleukin 1 and tumour necrosis factor alpha may be responsible for the lytic mechanism during anti-tumour antibody-dependent cell-mediated cytotoxicity. Brittish J Cancer 1995; 72:601-606.
33. Reali E, Spisani S, Gavioli R et al. IL-8 enhances antibody-dependent cellular cytotoxicity in human neutrophils. Immunol Cell Biol 1995; 73: 234-238
34. te Velde AA, de Waal Malefyt R, Huijbens RJF et al. IL-10 stimulates monocyte FcγR surface expression and cytotoxic activity: distinct regulation of ADCC by IFN-γ, IL-4 and IL-10. J Immunol 1992; 149:4048-4052.
31. Huang D, Zhou Y, Giscombe R et al. Dinucleotide repeat expansion in CTLA-4 gene leads to T cell hyperreactivity via the CD28 pathway in myasthenia gravis. J Neuroimm 2000; 105:69-77.
35. Huang D, ShiQin X, Zhou YH et al. Markers in the promotor region of Interleukin-10 (IL-10) gene in myasthenia gravis:Implications on diverse effects of IL-10 in the pathogenesis of the disease. J Neuroimm 1999; 94:82-87.
36. Huang D, Giscombe R, Matell G et al. Polymorphisms at -174 and in the 3′ flanking region of interleukin-6 (IL-6) gene in patients with myasthenia J Neuroimm 1999; 101:197-200.
37. Huang D, ShiQin X, Zhou YH et al. No evidence for Interleukin-4 gene conferring susceptibility to myasthenia gravis. J Neuroimm 1999; 92: 208-211.
38. Shaw G, Kamen R. A conserved AU sequence from the 3′untranslated region of GM-CSF mRNA mediates selective mRNA degradation. Cell 1986; 46:659-667.
39. Wang XB, Giscombe R, Qui Q et al. Abnormal expression of CTLA-4 in T cells from patients with myasthenia gravis is an effect of an AU-rich gene sequence. 2001; manuscript submitted.

Chapter 7

CTLA-4 in Multiple Sclerosis

Rebecca J. Greenwald, Yvette Latchman and Arlene H. Sharpe

Introduction

The B7:CD28/CTLA-4 pathway has a pivotal role in regulating T cell immune responses and manipulation of this key immunoregulatory pathway may lead to the development of therapeutic interventions to control autoimmunity. This pathway is complex because the B7-1 and CD86 costimulatory molecules have dual specificities for CD28 and CTLA-4. B7-1 and CD86, provide the major costimulatory signal for augmenting and sustaining T cell immune responses via interaction with CD28.[1] In contrast, engagement of B7 on antigen presenting cells (APCs) by CTLA-4 on T cells delivers signals that inhibit T cell responses.[2,3] Thus, ligation of CTLA-4 has the opposite effect on T cells as ligation of CD28, although both receptors bind the same B7 molecules on APCs. The important immunoregulatory roles of the B7:CD28/CTLA-4 pathway prompted studies investigating how this pathway influences the initiation and progression of autoimmune diseases. As key regulators of T cell activation, B7:CD28 and B7:CTLA-4 interactions play important roles during the pathogenesis of autoimmune diseases.

This chapter summarizes recent advances in our understanding of the role of the CD80/CD86:CD28/CTLA-4 pathway in the pathogenesis of autoimmune disease with emphasis on the role of CTLA-4 in regulating experimental autoimmune encephalomyelitis (EAE). EAE is a chronic inflammatory demyelinating disease of the central nervous system (CNS) and has been used as an animal model for human multiple sclerosis (MS). Disease susceptibility to EAE, diabetes, lupus, and autoimmune ovarian dysgenesis has been linked to the CD28/CTLA-4 locus.[4-8] Genetic studies indicate that these pathways may contribute to the pathogenesis of human autoimmune diseases. We will focus on studies in EAE models which provide insights into how CTLA-4 may regulate the pathogenesis of EAE and have relevance to understanding the role of CTLA-4 in MS. The field is complicated by the multiple models of EAE in which different genetic strains of mice are immunized with various myelin antigens. However, we will emphasize common themes that have emerged from these studies.

Receptors and Ligands in the B7:CD28/CTLA-4 Costimulatory Pathway

B7-1 and CD86 are expressed on the surface of B cells, dendritic cells, macrophages, and T cells.[9] Studies have shown distinct patterns of expression for B7-1 and CD86: CD86 is constitutively expressed at low levels on dendritic cells, macrophages, and B cells, and is rapidly upregulated in response to numerous activation stimuli, including cytokines, activation signals and infection.[1] B7-1 is upregulated later than CD86, suggesting that CD86 may be more important for initiating an immune response, whereas B7-1 may function to amplify and/or regulate an immune response. Both B7-1 and CD86 bind to CTLA-4 with at least a 10-fold higher affinity than they bind to CD28.[10] Their dissociation kinetics are also different. B7-1 has a slower "off" rate than CD86 which may account for the distinct functional roles of these molecules in different types of immune responses.[11] Whether B7-1 and CD86 have identical, overlapping, or distinct functions in vivo is an active area of research.

CD28 and CTLA-4 share a hexamer motif (MYPPPY) which contains the critical residues for binding to B7-1 and CD86.[11] They are closely linked on mouse chromosome 1. The expression patterns of CD28 and CTLA-4 on T cells are distinct. Whereas CD28 is expressed constitutively on T cells, CTLA-4 is upregulated upon activation with peak expression 24 to 48 hours after activation.[12]

B7:CD28/CTLA-4 Pathway and the Induction of EAE

The importance of the B7:CD28/CTLA-4 pathway in the induction of autoimmunity, and EAE, in particular was highlighted by studies with CTLA-4Ig, a fusion protein that blocks B7-1 and CD86 interactions with CD28 and CTLA-4.[13,14] When CTLA-4Ig is given at the time of immunization, the development of EAE can be prevented or ameliorated.[15-17] Likewise, treatment of myelin basic protein (MBP)-reactive lymph node cells with CTLA-4Ig in vitro prior to adoptive transfer in vivo reduced the severity of EAE in the recipients.[17] These findings demonstrated a key role for the B7:CD28/CTLA-4 costimulatory pathway in the generation of encephalitogenic T cells. However, no inhibitory effect on disease was observed after activated MBP-specific T cells were transferred into naive recipients and CTLA-4Ig was given to the recipients. Thus, these results suggested that B7-mediated costimulation is required for the induction, but not for the maintenance of T cell effector responses. Interestingly, administration of CTLA-4Ig prior to immunization resulted in the exacerbation of EAE.[16] The varied disease outcomes following CTLA-4Ig administration may depend on whether B7:CD28 or B7:CTLA-4 interactions were predominately blocked. Further insight into the role of the B7:CD28/CTLA-4 pathway has come from studies using antibodies and gene knockout mice.

B7:CD28 interactions in EAE have been addressed in several models of EAE using anti-B7 monoclonal antibodies (mAbs) and B7 gene deficient (-/-) mice. Similar to treatment with CTLA-4Ig, studies using anti-B7-1 and/or anti-CD86 mAbs have resulted in varying disease outcomes depending on the model system. However, these findings have indicated a distinct role for B7-1 and CD86 during the course of disease development in EAE. For instance, treatment with anti-B7-1 mAbs prevented EAE, while anti-CD86 mAbs either increased or had no effect on the severity of disease.[16,18,19] One possible explanation for these distinct effects on EAE induction is that B7-1 and CD86 exert distinct effects on Th1 versus Th2 differentiation.[18] Given the differences the kinetics of CD80 and CD86 ligand expression, CD86:CD28 interactions may predominate at the induction of the autoimmune response, whereas B7-1:CTLA-4 interactions may be more pronounced during the effector phase. In addition, potential differences in the intensity of signal via these receptors may also influence Th1 versus Th2 differentiation. Interestingly, CD80 is expressed highly in the CNS which may offer an explanation for its effectiveness in ameliorating EAE.[19, 20]

Although studies using CTLA-4Ig and anti-B7 mAbs have elucidated distinct roles for CD80 and CD86 during the course of EAE, potential pitfalls of using these reagents include: 1) Fc-receptor crosslinking, 2) anti-B7 mAbs may induce stimulatory signals through B7 expressed on T cells versus interactions of B7 on APCs with CD28 and CTLA-4 on T cells, 3) differences in affinity of CTLA-4Ig and anti-B7-1 or anti-CD86 for their respective targets, 4) differences in half-life and penetration of the fusion protein and antibody in vivo, and 5) antibody responses to the reagents.

In an attempt to clarify the antibody and CTLA-4Ig studies, mice deficient in CD80 and/or CD86 were generated. Distinct roles for B7-1 and CD86 were not observed in B7-deficient C57Bl/6 mice following immunization with myelin oligodendrocyte protein (MOG).[21] In fact, MOG-immunized B7-1$^{-/-}$ or CD86$^{-/-}$ C57Bl/6 mice developed EAE comparably to wild-type controls. On the SJL background, however, anti-B7-1 mAbs ameliorated disease, while anti-CD86 mAbs had no effect or increased disease severity.[16,18,19] The inconsistent disease phenotypes between anti-B7 mAbs and the B7 gene deficient mice may be due to variations in genetic background or distinct requirements for B7-1 or CD86 in different models of EAE. Although the gene-deficient studies support overlapping roles for B7-1 and CD86 in EAE, these studies have demonstrated that B7-1 and CD86-mediated costimulation are critical in the induction phase of EAE.

Similar to B7-1/CD86$^{-/-}$ mice, CD28$^{-/-}$ C57BL/6 mice are resistant to EAE induction by MOG 35-55 immunization and inflammatory infiltrates in the CNS are limited to the meninges.[21,22] Likewise, the development of spontaneous EAE in RAG2$^{-/-}$ TCR transgenic mice specific for MBP Ac1-11 was prevented in CD28$^{-/-}$ RAG2$^{-/-}$ MBP TCR transgenic mice.[23] Taken together, these studies demonstrate a critical role for B7:CD28 interactions in the induction of EAE.

B7:CD28 Costimulation in the Effector Phase of Autoimmune Disease

While important for the induction, expansion, and differentiation of naïve T cells, the role of B7:CD28 interactions in the activation of effector T cells and sustaining inflammatory immune responses in the target organ has been appreciated only recently. This is a critical issue for clinical autoimmune disease with important therapeutic implications, since at the time of clinical presentation, autoreactive T cells are already generated and expanded.

A key role for B7 costimulation in the effector phase was demonstrated by adoptive transfer experiments using B7-1/CD86$^{-/-}$ mice.[21] When wild-type MOG-specific T cells were transferred into B7-1/CD86$^{-/-}$ recipients, transient disease developed with markedly reduced severity, and the mice recovered from disease. In these animals, inflammatory Th1 cytokine production was elevated, however the number of inflammatory lesions in the CNS was markedly reduced and restricted to the meninges. These findings suggest that B7-mediated costimulation is critical for either entry into the CNS parenchyma or to sustain T cell responses in the CNS parenchyma. Studies with blocking anti-CD28 mAbs and CTLA-4Ig confirm a role for B7:CD28 interactions in sustaining inflammation during the effector phase of EAE.

The Role of B7:CTLA-4 Interactions in EAE

The fatal lymphoproliferative disease that develops in CTLA-4-deficient mice provided direct evidence for a critical role for CTLA-4 in down-regulating T cell activation.[2]

This phenotype also suggested that B7:CTLA-4 interactions may be required for inducing and/or maintaining peripheral T cell tolerance and preventing autoimmunity. In several animal models, treatment with anti-CTLA-4 mAbs exacerbates EAE induced by adoptive transfer of primed cells or active immunization with myelin antigens.[24,25] The increase in disease severity following CTLA-4 blockade suggests that CTLA-4 functions under normal conditions to control autoimmune responses. When anti-CTLA-4 mAbs are given at the onset of clinical EAE, increased mortality is observed.[26] Similarly, in a model of relapsing-remitting EAE, when anti-CTLA-4 mAbs are given during remission, the severity of the relapse is markedly increased.[24] CTLA-4 also regulates epitope spreading, a process that occurs when immune responses to autoantigens become more diverse as the response progresses.[27] Not only does CTLA-4 blockade augment responses to the specific encephalitogenic epitope, but also to other related epitopes. Thus, CTLA-4 appears to have a key role in the down-regulation of the autoimmune response and the regulation of EAE.

Recent studies have demonstrated a role for CTLA-4 in the induction of peripheral T cell tolerance using naive TCR transgenic T cells lacking CTLA-4.[28] Following encounter with a tolerogenic stimulus in vivo and restimulation in vitro, wild-type TCR trangenic T cells exhibit defects in proliferation, cytokine production, and cell cycle progression. In marked contrast, CTLA-4-deficient TCR transgenic T cells proliferate, secrete IL-2, and progress through the cell cycle following administration of a tolerogenic stimulus. Interestingly, this tolerance induction is not associated with a defect in activation-induced cell death, but related to the role for CTLA-4 in regulating cell cycle progression. Further studies are required to investigate the role of CTLA-4 in maintaining tolerance in EAE.

The role for CTLA-4 in activation and tolerance has lead to important insights into how CTLA-4 may regulate T cell responses to self-antigens. CTLA-4 may raise the threshold for activation and reduce the potential for activation by weak signals, including to self-antigens. As T cells become activated, CTLA-4 limits the immune response, thereby maintaining immunologic homeostasis. Thus, CTLA-4 is likely to prevent stimulation of T cells by weakly reactive self-antigens.

Some studies have suggested that CTLA-4 may alter T cell function by regulating the production of soluble immunoregulatory cytokines such as transforming growth factor β (TGF-β).[29] Both TGF-β and CTLA-4 down-regulate immune responses, and the phenotype of the $TGF\beta^{-/-}$ and $CTLA\text{-}4^{-/-}$ mouse strains are similar. Further support for a role for CTLA-4 in controlling production of soluble regulatory cytokines comes from a recent study which suggests that the defect in $CTLA\text{-}4^{-/-}$ mice was not cell autonomous.[30] Reconstitution of $RAG2^{-/-}$ mice with a mixture of wild-type and $CTLA\text{-}4^{-/-}$ bone marrow did not result in disease, whereas $RAG2^{-/-}$ recipients of $CTLA\text{-}4^{-/-}$ bone marrow developed lymphoproliferative infiltration into the heart and liver, but not the lymphoproliferative disorder observed in the $CTLA\text{-}4^{-/-}$ animals. In these studies, however, a role for TGF-β was not supported, but other soluble factors may be involved.

In contrast, other studies suggest that the defect exhibited in $CTLA\text{-}4^{-/-}$ T cells is cell intrinsic. CTLA-4 may directly influence TCR-mediated signals. Studies of CTLA-4 signaling have indicated that following T cell activation CTLA-4 may block tyrosine phosphorylation by directly interacting with the TCR complex z chain.[31] B7-1:CTLA-4 interactions also have been linked with antagonism of TCR-mediated signals in a CD28-independent system.[32] In addition, B7:CTLA-4 interactions also may interfere with CD28-mediated signals. Since CTLA-4 has a higher affinity for the B7 ligands, as expression of CTLA-4 increases during the progression of the immune response, CTLA-4 may outcompete CD28 for binding to the B7 ligands and/or shared downstream signaling molecules. Indirect evidence for CTLA-4 directly antagonizing CD28 signaling is provided by the finding that both CD28 and CTLA-4 can associate with the phosphatidylinositol 3-kinase[33] and serine-threonine phosphatase PP2A.[34] A competition mechanism also is supported by the crystal structures of B7[35] and CTLA-4.[36] CTLA-4 dimerization is necessary for the formation of high avidity complexes with its B7 ligands and transmission of downregulatory signals. The dimerization properties of CTLA-4 are unusual in that each CTLA-4 dimer can bind two independent B7 bivalent B7 molecules. These structural properties suggest that there is a periodic arrangement of B7 ligands and CTLA-4 within the immunological synapse, and this may promote the recruitment of inhibitory signaling molecules through interaction with either the cytoplasmic or extracellular domains in CTLA-4 and B7.

Recent reports have demonstrated constitutive expression of CTLA-4 on $CD4^+$ $CD25^+$ T regulatory cells and suggest that CTLA-4 may be critical for their function.[37,38] The function of CTLA-4 on T regulatory cells in still unclear. Similar to the role for CTLA-4 in T cell activation and differentiation, studies have indicated that CTLA-4 on T regulatory cells may produce immunoregulatory cytokines such as IL-10, TGF-β, or other immunosuppressive cytokines. Understanding the molecular mechanisms of CTLA-4 function is an active area of investigation and further studies are required to clarify the role of CTLA-4 in autoimmunity.

The outcome of an immune response involves a balance between CD28-mediated T cell activation and CTLA-4-mediated inhibition. How signals through CD28 and CTLA-4 are coordinated is not clear and is an area of current investigation. CTLA-4 may inhibit T cell responses by outcompeting CD28 for binding to B7, by antagonizing CD28-mediated signaling and/or antagonizing TCR-mediated signaling directly. These mechanisms are not mutually exclusive. As discussed above, there are data to support all three of these mechanisms.

New Pathways within the B7:CD28 Superfamily and EAE

The emergence of two novel pathways within the B7:CD28/CTLA-4 superfamily gives impetus to analyzing their role in EAE. The first pathway involves the interaction of the CD28 homolog inducible co-stimulator (ICOS) and its ligand (B7h,B7RP-1,GL50).[39-42] ICOS is expressed on T cells after activation. ICOS appears to be important for the augmentation and differentiation of Th2 cells.[43,44] Mice deficient in ICOS have impaired immunoglobulin class switching and germinal centre formation, and T helper cell differentiation.[45-47] Furthermore, $ICOS^{-/-}$ mice on the C57BL/6 x129F_2 background immunized with MOG peptide exhibit enhanced susceptibility to EAE with massive infiltration in the brain and spinal cord.[47] Wild-type mice on the same background exhibited mild disease. Splenocytes from $ICOS^{+/+}$ activated with MOG peptide produced substantial levels of IL-13, while $ICOS^{-/-}$ mice produced no detectable IL-13. This study suggests a role for ICOS in regulating the development/or progression of EAE. The ICOS gene is closely linked to the CD28 and CTLA-4 genes. Therefore, genetic studies linking the CD28/CTLA-4 locus to autoimmune diseases also may involve ICOS. This is particularly intriguing in light of the immunoregulatory role of ICOS in EAE. Further studies are needed to understand the role of ICOS during the induction and effector phases of EAE.

The second pathway involves the interaction of the programmed death-1 (PD-1) molecule with its newly discovered ligands PD-L1 and PD-L2.[48-51] PD-1 is a CD28 homologue which contains an immunoreceptor tyrosine-based inhibitory motif (ITIM) in its cytoplasmic tail. PD-1 is expressed on a subset of thymocytes and is upregulated on T cells, B cells and myeloid cells after their activation.[52,53] The mRNA for PD-1 ligands is found in a variety of normal tissues and in activated antigen presenting cells.[49,51] C57BL/6 mice that lack PD-1 developed a lupus-like arthritis and glomerulonephritis and showed increased levels of serum IgG3 in vivo and augmented B cell proliferative responses to anti-IgM in vitro.[54] Furthermore, PD-$1^{-/-}$ mice crossed with the H-$2L^d$-specific 2C-TCR transgenic mice in the H-$2^{b/d}$ background developed a chronic graft–versus host-like disease. In contrast PD-$1^{+/+}$ crossed on the same background with 2C-TCR showed no evidence of disease.[49] Balb/c PD-$1^{-/-}$ mice also have been shown to develop a fatal dilated cardiomyopathy with massive thrombosis.[55] The expression of PD-1 ligands in non-lymphoid tissues suggests suggests a previously unknown means by which self reactive T cells that escape into the periphery may be prevented from causing tissue injury, and suggests that the PD-L:PD-1 pathway may be critical for the induction and/or maintenance of peripheral T cell tolerance to self antigens. Further studies will have to be undertaken to evaluate the role of this inhibitory pathway in regulating EAE. In addition, studies of the interaction of the B7:CTLA-4 and PD-L:PD-1 pathways may clarify whether these pathways have complementary roles in the induction and/or maintenance of tolerance.

Concluding Remarks

The B7:CD28/CTLA-4 pathway has critical roles in regulating T cell activation and tolerance. Blocking B7 costimulation in vivo can prevent or delay the initiation and progression of autoimmune diseases. As a result, this pathway represents an important therapeutic target for controlling human autoimmune diseases. CD28 and CTLA-4 have opposing effects on T cell activation. CD28 lowers the threshold for effective TCR activation, and may enhance lower avidity interactions during autoreactive T cell responses. In contrast, CTLA-4 downregulates T cell activation, and may prevent T cell activation by low strength TCR signals and protect against autoimmune responses. Especially encouraging for therapeutic intervention involving the B7:CD28/CTLA-4 pathway is the recent demonstration that B7-CD28 interactions are required during the effector phases of EAE. However, because B7 blockade can also result in increased T cell proliferation by inhibiting CTLA-4 signaling, further studies are needed to understand how to most effectively manipulate this pathway to control human autoimmune diseases. The recent discovery of two new pathways within the B7:CD28 superfamily has revealed new means by which the activation of T cells is regulated. These new pathways appear particularly important for regulating the responses of activated T cells and may provide novel therapeutic targets for manipulating effector T cell responses.

References

1. McAdam AJ, Schweitzer AN, Sharpe AH. The role of B7 co-stimulation in activation and differentiation of CD4+ and CD8+ T cells. Immunol Rev 1998; 165:231.
2. Tivol E, Borriello F, Schweitzer A et al. Loss of CTLA-4 leads to massive lymphoproliferation and fatal multiorgan destruction, revealing a critical negative regulatory role of CTLA-4. Immunity 1995; 3:541.
3. Waterhouse P, Penninger J, Timms E et al. Griesser, and T. Mak. Lymphoproliferative disorders with early lethality in mice deficient in CTLA-4. Science 1995; 270:985.

4. Encinas JA, Weiner HL, KuchrooVK. Inheritance of susceptibility to experimental autoimmune encephalomyelitis. J Neurosci Res 1996; 45:655.
5. Encinas JA, Lees MB, Sobel RA et al. Genetic analysis of susceptibility to experimental autoimmune encephalomyelitis in a cross between SJL/J and B10.S mice. J Immunol 1996; 157:2186.
6. Griffiths MM, Encinas JA, Remmers EF et al. Mapping autoimmunity genes. Curr Opin Immunol 1999; 11:689.
7. Lyons PA, Wicker LS. Localising Quantitative Trait Loci in the NOD mouse Model of Type 1 Diabetes. Basel: Karger, 1999.
8. Hill NJ, Lyons PA, Armitage N et al. NOD Idd5 locus controls insulitis and diabetes and overlaps the orthologous CTLA4/IDDM12 and NRAMP1 loci in humans. Diabetes 2000; 49:1744.
9. Hathcock KS, Laszlo G, Pucillo C et al. Comparative analysis of B7-1 and CD86 costimulatory ligands: expression and function. J Exp Med 1994; 180:631.
10. van der Merwe PA, Bodian DL, Daenke S et al. CD80 (B7-1) binds both CD28 and CTLA-4 with a low affinity and very fast kinetics. J Exp Med 1997; 185:393.
11. Linsley PS, Greene JL, Brady W et al. Human B7-1 (CD80) and CD86 (CD86) bind with similar avidities but distinct kinetics to CD28 and CTLA4 receptors. Immunity 1994; 1:793.
12. Linsley P, Bradshaw J, Greene J et al. Intracellular trafficking of CTLA4 and focal localization towards sites of TcR engagement. Immunity 1996; 4:535.
13. Linsley PS, Wallace PM, Johnson J et al. Immunosuppression in vivo by a soluble form of the CTLA-4 T cell activation molecule. Science 1992; 257:792.
14. Lenschow D, Zeng Y, Thistlewaite J et al. Longterm survival of Xenogenic Panceartic Islet grafts induced by CTLA4-Ig. Science 1992; 257:789.
15. Cross AH, Girard TJ, Giacoletto KS et al. Long-term inhibition of murine experimental autoimmune encephalomyelitis using CTLA4-Fc supports a key role for CD28 costimulation. Journal of Clinical Investigation 1995; 95:2783.
16. Racke M, Scott D, Quigley L et al. Distinct roles of B7-1 (CD-80) and CD86 (CD-86) in the initiation of Experimental Allergic Encephalomyelitis. J. Clinical Investigation 1995; 96:2195.
17. Perrin PJ, Scott D, Quigley L et al. Role of B7/CD28 CTLA-4 in the induction of chronic relapsing experimental allergic encephalomyelitis. Journal of Immunology 1995; 154:1481.
18. Kuchroo V, Prabhu Das M, Brown JA et al. B7-1 and CD86 costimulatory molecules differentially activate the Th1/Th2 developmental pathways: application to autoimmune disease therapy. Cell 1995; 80:707.
19. Miller S, Vanderlugt C, Lenschow D et al. Blockade of CD28-B7-1 interaction prevents epitope spreading and clinical relapses of murine EAE. Immunity 1995; 3:739.
20. Karandikar NJ, Vanderlugt CL, Eagar T et al. Tissue-specific up-regulation of B7-1 expression and function during the course of murine relapsing experimental autoimmune encephalomyelitis. J Immunol 1998; 161:192.
21. Chang TC, Jabs C, Sobel RA et al. Studies in B7-deficient mice reveal a critical role for B7 costimulation in both the initiation and effector phases of. Journal of Experimental Medicine 1999; 190:733.
22. Perrin PJ, Lavi E, Rumbley CA et al. Experimental autoimmune meningitis: a novel neurological disease in CD28-deficient mice. Clin Immunol 1999; 91:41.
23. Oliveira-dos-Santos AJ, Ho A, Tada Y et al. CD28 costimulation is crucial for the development of spontaneous autoimmune encephalomyelitis. J Immunol 1999; 162:4490.
24. Karandikar N, Vanderlugt C, Walnus T et al. CTLA-4-a negative regulator of autoimmune disease. Journal of Experimental Medicine 1996; 184:783.
25. Hurwitz AA, Sullivan TJ, Krummel MF et al. Specific blockade of CTLA-4/B7 interactions results in exacerbated clinical and histologic disease in an actively-induced model of experimental allergic encephalomyelitis. J Neuroimmunol 1997; 73:57.
26. Perrin PJ, Maldonado JH, Davis TA et al. CTLA-4 blockade enhances clinical disease and cytokine production during experimental allergic encephalomyelitis. J Immunol 1996; 157:1333.
27. Karandikar NJ, Vanderlugt CL, Bluestone JA et al. Targeting the B7/CD28:CTLA-4 costimulatory system in CNS autoimmune disease. J Neuroimmunol 1998; 89:10.
28. Greenwald RJ, Boussiotis VA, Lorsbach RB et al. CTLA-4 regulates induction of anergy in vivo. Immunity 2001; 14:144.
29. Chen W, Jin W, Wahl SM. Engagement of cytotoxic T lymphocyte-associated antigen 4 (CTLA-4) induces transforming growth factor beta (TGF-beta) production by murine CD4(+) T cells. J Exp Med 1998; 188:1849.

30. Bachmann MF, Kohler G, Ecabert B et al. Cutting edge: lymphoproliferative disease in the absence of CTLA-4 is not T cell autonomous. J Immunol 1999; 163:1128.
31. Lee KM, Chuang E, Griffin M et al. Molecular basis of T cell inactivation by CTLA-4. Science 1998; 282:2263.
32. Fallarino F, Fields PE, Gajewski TF. B7-1 engagement of cytotoxic T lymphocyte antigen 4 inhibits T cell activation in the absence of CD28. J Exp Med 1998; 188:205.
33. Schneider H, Prasad KV, Shoelson SE et al. CTLA-4 binding to the lipid kinase phosphatidylinositol 3-kinase in T cells. J Exp Med 1995; 181:351.
34. Chuang E, Fisher TS, Morgan RW et al. The CD28 and CTLA-4 receptors associate with the serine/threonine phosphatase PP2A. Immunity 2000; 13:313.
35. Ikemizu S, Gilbert RJ, Fennelly JA et al. Structure and dimerization of a soluble form of B7-1. Immunity 2000; 12:51.
36. Ostrov DA, Shi W, Schwartz JC et al. Structure of murine CTLA-4 and its role in modulating T cell responsiveness. Science 2000; 290:816.
37. Read S, Malmstrom V, Powrie F. Cytotoxic T lymphocyte-associated antigen 4 plays an essential role in the function of CD25(+)CD4(+) regulatory cells that control intestinal inflammation. J Exp Med 2000; 192:295.
38. Takahashi T, Tagami T, Yamazaki S et al. Immunologic self-tolerance maintained by CD25(+)CD4(+) regulatory T cells constitutively expressing cytotoxic T lymphocyte-associated antigen 4. J Exp Med 2000; 192:303.
39. Hutloff A, Dittrich AM, Beier KC et al. ICOS is an inducible T-cell co-stimulator structurally and functionally related to CD28. Nature 1999; 397:263.
40. Wang S, Zhu G, Chapoval AI et al. Costimulation of T cells by B7-H2, a B7-like molecule that binds ICOS [In Process Citation]. Blood 2000; 96:2808.
41. Yoshinaga SK, Zhang M, Pistillo J et al. Characterization of a new human B7-related protein: B7RP-1 is the ligand to the co-stimulatory protein ICOS [In Process Citation]. Int Immunol 2000; 12:1439.
42. Ling V, Wu W, Finnerty HF et al. Cutting edge: identification of GL50, a novel B7-like protein that functionally binds to ICOS receptor. J Immunol 2000; 164:1653.
43. McAdam AJ, Chang TT, Lumelsky AE et al. Mouse inducible costimulatory molecule (ICOS) expression is enhanced by CD28 costimulation and regulates differentiation of CD4(+) T cells [In Process Citation]. J Immunol 2000; 165:5035.
44. Coyle AJ, Lehar S, Lloyd C et al. The CD28-related molecule ICOS is required for effective T cell- dependent immune responses. Immunity 2000; 13:95.
45. McAdam AJ, Greenwald RJ, Levin MA et al. ICOS is critical for CD40-mediated antibody class switching. Nature 2001; 409:102.
46. Tafuri A, Shahinian A, Friedheim B et al. ICOS is essential for effective T-helper -cell responses. Nature 2001; 409:105.
47. Dong C, Juedes AE, Temann UA et al. ICOS co-stimulatory receptor is essential for T-cell activation and function. Nature 2001; 409:97.
48. Ishida Y, Agata Y, Shibahara K et al. Induced expression of PD-1, a novel member of the immunoglobulin gene superfamily, upon programmed cell death. Embo J 1992; 11:3887.
49. Freeman GJ, Long AJ, Iwai Y et al. Engagement of the PD-1 immunoinhibitory receptor by a novel B7 family member leads to negative regulation of lymphocyte activation. J Exp Med 2000; 192:1027.
50. Dong H, Zhu G, Tamada K et al. B7-H1, a third member of the B7 family, co-stimulates T-cell proliferation and interleukin-10 secretion. Nat Med 1999; 5:1365.
51. Latchman Y, Wood CR, Chernova T et al. PD-L2 is a second ligand for PD-I and inhibits T cell activation. Nat Immunol 2001; 2:261.
52. Nishimura H, Agata Y, Kawasaki A et al. Developmentally regulated expression of the PD-1 protein on the surface of double-negative (CD4-CD8-) thymocytes. Int Immunol 1996; 8:773.
53. Agata Y, Kawasaki A, Nishimura H et al. Expression of the PD-1 antigen on the surface of stimulated mouse T and B lymphocytes. Int Immunol 1996; 8:765.
54. Nishimura H, Minato N, Nakano T et al. Immunological studies on PD-1 deficient mice: implication of PD-1 as a negative regulator for B cell responses. Int Immunol 1998; 10:1563.
55. Nishimura H, Okazaki T, Tanaka Y et al. Autoimmune dilated cardiomyopathy in PD-1 receptor-deficient mice. Science 2001; 291:319.

Chapter 8

CTLA-4 in Addison's Disease

Klaus Badenhoop

Introduction

Addison's disease is a rare autoimmune disorder of adrenal destruction leading to death if unrecognised and untreated. Usually non-surgical adrenal insufficiency is caused by either tuberculous granuloma, other infectious agents such as observed in AIDS or has no clear etiology in about 70-80% that is considered to be idiopathic. In this form of adrenal dysfunction the adrenal glands are atrophic and contain only few cortical cells. The adrenal medulla is usually spared from the destruction. In contrast the tuberculous adrenal disease leads to granuloma and caseation with subsequent calcification.

Idiopathic adrenal insufficiency is in most cases due to autoimmune destruction of adrenocortical cells: antibodies against adrenal cells and the adrenal autoantigen steroid-21-hydroxylase are detected in the majority (80%) of newly manifested patients.[1] A recently developed radiobinding assay has been reported to be highly sensitive and specific for Addison's disease.[2]

As in other autoimmune endocrinopathies, there is a genetic susceptibility conferred by several gene loci. Immune destruction of adrenal cells is marked by an infiltration of autoreactive T-cells and the appearance of antibodies directed against intracellular enzymes such as 21- and 17-α-hydroxylases. Patients may present with mild weakness and malaise that—if underestimated—would become a problem of medical neglicence that may even happen in precomatous patients. Once the diagnosis is confirmed by low cortisol levels after ACTH stimulation, patients are treated with life-long glucocorticoid and mineralocorticoid substitution therapy. Since this disease is still underrecognised, exact figures of incidence and prevalence are lacking. However in Italy and in the UK such calculations have been performed: Laureti et al have calculated the prevalence of Addison's disease to be in the order of 117 per million in Umbria,[3] whereas Kong and Jeffcoate had previously reported 110 per million in Nottingham UK with an incidence of 5.6 per million.[4] Through affected patients one can identify antibody positive relatives with still normal adrenal function that are presymptomatic individuals at high risk for Addison's disease.

Addison's Disease as Part of a Polyglandular Syndrome

Addison's disease may occur as an isolated adrenal disorder or in concomittant or subsequent dysfunction of other hormone secreting glands or autoimmune syndromes named as polyglandular deficiency. Polyglandular deficiency type 1 is an autosomal recessive condition with adrenocortical failure, hypoparathyroidism, mucocutaneous candidiasis

CTLA-4 in Autoimmune Disease, edited by Flemming Pociot. ©2004 Eurekah.com.

and other conditions that affects usually children. The synonymous autoimmune polyendocrinopathy-candidiasis-ectodermal dystrophy (APECED) is rare and occurs more frequently in genetically isolated populations such as the Finnish (1/25000), the Sardic (1/14500) or the Iranian Jews (1/9000). The gene defect causing this disorder has been identified using the positional cloning approach and resides on chromosome 21q22.3. The gene has been named AIRE-1 for autoimmune regulator and several mutations have been detected in patients. Although the gene structure has been characterised, its exact function needs to be determined. However such AIRE-1 mutations—at least the common ones—are not found among patients with isolated Addison's disease.[5]

The adrenal insufficiency of polyglandular deficiency type 2 is not linked to the AIRE-1 gene but occcurs predominantly with thyroid autoimmnity, vitiligo, primary hypogonadism or type 1 diabetes. This syndrome is associated with the HLA-DR3 specificity, similar to isolated Addison's disease. Besides the shared genetic susceptibility of HLA-DR3 there is also the allele HLA-DQA1*0501 that is common to all endocrine autoimmune diseases: Graves′ disease, Hashimoto′s and post-partum thyroiditis, type 1 diabetes and Addison's disease.[6] Some of these patients may also have detectable antibodies to 21-hydroxylase and about a third of these individuals will develop Addison's disease.

Thus individuals at risk to contract an endocrine autoimmune disorder such as Addison's disease can be detected by both antibody testing and genetic risk markers of the HLA DR and DQ genes. Amongst other genes that confer susceptibility to type 1 diabetes, Graves′ disease and also Addison's disease is the cytotoxic T-lymphocyte associated antigen-4 (CTLA-4). This gene codes for a protein that is known as a negative regulator of T cell activation and cytokine production. The immunobiology of CTLA-4 has been discussed in detail in this book.

The CTLA4 Gene and Addison's Disease

On chromosome 2q33 in man a susceptibility gene has been mapped for type 1 diabetes (IDDM) and named IDDM12.[7] This region contains the CTLA4 gene (cytotoxic T lymphocyte antigen 4) that codes for a receptor structure on T cells interacting with the B7 accessory molecules thus forming a costimulatory complex. The B7:CD28/CTLA4 costimulatory pathway consists of two costimulatory ligands on antigen-presenting cells. CTLA4 is only expressed on activated T-cells. B7-1 (CD80) and B7-2 (CD86) have a higher affinity to CTLA4 than to CD28, which is also expressed by resting T-cells. CTLA4 may thus be regarded as the predominant B7 receptor on activated T-cells and a key regulatory element in the interaction with antigen presenting cells (rev. in ref. 8). Also, CTLA4 has been shown to mediate antigen specific apoptosis and therefore it represents a negative regulator for T cell function.[9] In a murine model of autoimmune encephalomyelitis CTLA4 mediates the downregulation of the immune response.[10] Since it is generally believed that T cells are the main effector cells in endocrine autoimmunity,[11,12] CTLA4 is a prime candidate locus for susceptibility. Whether CTLA4 may serve as an additional, complementary risk marker in Addison's disease was addressed in two studies in recent years.

Subjects

Patients with Addison's disease (n=76) were analysed and compared with 466 controls. We also analysed 73 patients with Hashimoto′s thyroiditis, which was diagnosed by the presence of goitre, hypothyroidism, and elevated microsomal or thyroid peroxidase autoantibodies. Thyroid ultrasound showed typically a reduced echogenicity compatible with thyroid autoimmunity. The group of patients included patients from our previous report[6] from Frankfurt, Germany, and Toronto, Canada (n=18). This patient group was included since Hashimoto's thyroiditis comprises the most common autoimmune condition that is present in patients with other immune-mediated diseases affecting the endocrine but also other organs systems.

Addison's disease was diagnosed by primary adrenocortical insufficiency. There was no evidence for tuberculosis or adrenoleucodystrophy. The age of disease onset varied from 16 to 42 years and no neurological deficits could be detected. In 26 patients thyroid (Graves' disease or Hashimoto's thyroiditis) or β-cell (IDDM) autoimmune disease was also present. Patients were from Frankfurt/Main, Mannheim or Berlin, Germany.

Healthy controls were randomly collected from Frankfurt/Main, Mannheim or Berlin, Germany, (n=383) and Toronto, Canada (n=83). There was no family history of type 1 diabetes, Graves' disease, Hashimoto's thyroiditis, or Addison's disease. The distribution of CTLA4 alleles did not differ between controls from Canada or from Germany.

Methods

The CTLA4 exon 1 position 49 (codon 17) polymorphism was studied by a combination of methods (PCR and SSCP) as described previously.[13] Briefly, PCR was performed with oligonucleotides forward 5'-GCTCTACTTCCTGAAGACCT-3' and reverse 5'-AGTCTCACTCACCTTTGCAG-3', designed according to the published human CTLA4-cDNA sequence (14) using 0.2μg genomic DNA, 1U Taq Polymerase (Gibco BRL), 20pmol each primer and 8mmol dNTP's under the following conditions: initial denaturation for 4 min at 94°C, annealing for 45 sec at 58°C, extension for 45 sec at 72°C, denaturation for 45 sec at 94°C (30 cycles) and a final extension for 4 min at 72°C.

Single Strand Conformation Polymorphism (SSCP) Analysis of CTLA4 Polymorphisms

PCR products were analysed for variants by SSCP. Aliquots of the PCR product (2μl) were mixed with 2.3μl deionized formamide, incubated for 5 min at 95°C and loaded onto an 8% polyacrylamide gel. Gel electrophoresis was carried out at 10mA (10W, max. 1000V) for 2.5hrs keeping constantly 8°C on a Multiphor II apparatus and a Multitemp cooling system (LKB Pharmacia, Freiburg, Germany). Silverstaining of the gels was used to detect variant conformational fragments. These conformational variants corresponded to nucleotide substitutions which was confirmed by restriction enzyme analysis using BbvI: it defined a G at position 49 (88/74 bp fragments) or an A (no digestion of the 162 bp fragment). DNA fragments were resolved in 2.0% agarose gels stained with SYBR Green I (Molecular Probes, Leiden, Netherlands).

Typing for HLA DQA1 and DQB1 Alleles

Patients with Hashimoto's thyroiditis (n=66), Addison's disease (n=56) and controls (n=230) were typed for HLA DQA1 and DQB1 alleles as previously described.[6]

Statistical Methods

Patients and controls that were defined positive for an allele (phenotypic allele frequencies) were compared by the chi square test with Yates' correction and Fisher's exact test where appropriate (one number < 5). P-values had to be multiplied by the numbers of alleles tested (p_{corr}). Statistical significance was defined at $p<0.05$. Relative risks (RR) were calculated with Woolf's formula.

CTLA4 Exon 1 Polymorphisms in Patients with Addison's Disease

Eighteen percent of patients with Addison's disease were homozygous for Ala, 52% were heterozygous (Ala/Thr) and 30% were homozygous for Thr. The patients with Addison's disease and other autoimmune endocrine disorders did not differ from the controls or the whole group.

Patients with Addison's disease selected for the presence of HLA DQA1*0501 (n=53, 71% of all HLA DQA1 typed patients) were significantly more positive for the Ala allele: 40 (75%) patients compared with 59 (58%) DQA1*0501+ controls, $p<0.05$. Also the gene frequencies of Ala were borderline significantly higher in DQA1*0501+ patients compared with controls selected for the same DQA1 allele ($p=0.05$), Table 2.

CTLA4 Exon 1 Polymorphisms in Patients with Autoimmune Hypothyroidism (Hashimoto's Thyroiditis)

Significantly more patients were homozygous for Ala (22% vs. 15%) or heterozygous for Ala/Thr (53% vs. 46%) and less patients homozygous for Thr (25% vs. 39%), ($p<0.04$, Table 1). The gene frequency of Ala was higher in patients (49%) than in controls (38%, $p<0.02$). Furthermore, the Ala phenotype was more frequent in patients (75%) than in controls (61%, $p<0.03$, Table 1).

Patients positive for the risk marker HLA DQA1*0501 or negative did not differ from controls carrying the same HLA DQ alleles for the CTLA4 dimorphism (data not shown). Canadian patients did not differ from Germans for CTLA4 or HLA DQA1 alleles.

The distribution of CTLA4 alleles in all studied groups were in Hardy-Weinberg equilibrium, i.e., observed and expected figures did not differ.

Discussion

The CTLA4 gene has been implicated in several endocrine autoimmune disorders. The CTLA4 Ala^{17} is associated with IDDM and Graves'disease, whereas linkage was observed for IDDM (7,13). Since this CTLA codon 17 polymorphism is only diallelic it is less sensitive in association or linkage studies. A closely linked microsatellite polymorphism of the CTLA4 gene has a variability of 21 alleles and is situated in the 3'-untranslated region as a $(AT)_n$ repeat. The 106 bp allele of this microsatellite shows a particular association with Graves' disease both in Japan[15] and in Great Britain.[16] The latter report also finds this allele increased in patients with autoimmune hypothyroidism due to Hashimoto's thyroiditis.

We extend our recent report of the CTLA4 codon 17 dimorphism[13] to Hashimoto's thyroiditis, where 75% of patients have at least one Ala containing allele. The presence of particular HLA DQ alleles does not affect this association. In contrast, patients with Addison's disease and the predisposing HLA DQA1*0501 carry significantly more often at least one CTLA4 Ala^{17} allele.

This suggests that susceptibility to Addison's disease may, at least in a subgroup defined by HLA DQ risk alleles, be influenced by CTLA4 genotypes. A similar interaction

Table 1. CTLA4 exon 1 polymorphism in patients with Hashimoto's thyroiditis (HT), Addison's disease (AD) and in controls (Con)

Nucleotides at Position 49 (Aminoacids at Position 17)	HT n=73(%)	AD n=76(%)	Con n=466(%)
GG (Ala/Ala)	16(22)[a]	14(18)	68(15)
AG (Thr/Ala)	39(53)[a]	39(52)	215(46)
AA (Thr/Thr)	18(25)[a]	23(30)	183(39)
	[a]$p<0.04$		
Gene frequencies:			
G(Ala)	71(49)[b]	67(44)	351(38)
A(Thr)	75(51)[b]	85(56)	581(62)
	[b]$p<0.02$		
Phenotype frequencies:			
G(Ala)+	55(75)[c]	53(70)	283(61)
	[c]$p<0.03$		

P-values are given for the chi-square tests calculated on the 3x2 or 2x2 tables, comparing patients with HT (indicated under the columns) or Ad with controls, respectively.

between HLA DRB1*04 genotype and the CTLA4 Ala[17] allele has been observed by us in patients with rheumatoid arthritis.[24]

Our genetic findings may reflect differences between individuals who carry either the alanine or threonine CTLA4 codon 17 allele with respect to their macrophage activation or T-cell stimulation capabilities. This may relate to current concepts of the immune pathogenesis of those disorders.

Macrophage activation subsequent to a T-helper 1 response is thought to mediate organ specific autoimmunity in IDDM, whereas the antibody formation in Graves' disease is believed to result from a T-helper 2 action, rev. in.[12] Cytokine profiles of thyroid tissue derived from Graves' disease patients have been reported to show a pattern consistent with a T helper type 2 response, i.e., an increase of interleukin 4 and interleukin 10 levels.[17]

Since T-lymphocytes are thought to be the prime mediators of thyroid and also adrenal autoimmunity, the CTLA4 phenotype may affect T-cell function in the pathogenesis of both thyroid autoimmunity and Addison's disease. B7, the natural ligand of the CTLA4 molecule, is not expressed by thyrocytes, but by antigen presenting cells within the thyroid. Intrathyroidal macrophages have a higher density of CD86 on their surfaces compared with peripheral monocytes.[18] This makes it likely, that thyroidal cofactors lead to a higher expression level of CD86.

Although the exon 1 alanine/threonine substitution of the CTLA4 gene is not known to be of functional relevance, this polymorphism may be linked to the $(AT)_n$ microsatellite, that is situated in the 3' untranslated region and could affect RNA stability.[15]

Further studying the expression of this gene will provide more evidence how antigen polymorphism and immune dysregulation contribute to endocrine autoimmunity.

In our study we observed no overall significant difference between patients with Addison's disease and controls, although both gene and phenotype frequencies of the CTLA4 ala 17 allele were higher among the patients. However we did observe a significantly higher gene and phenotype frequency in patients stratified for the high risk HLA DQA1*0501

*Table 2. CTLA4 exon 1 polymorphism in patients with Addison's disease (AD) and controls (Con) analysed with respect to the presence of HLA DQA1*0301 and DQA1*0501*

	AD GG Ala/Ala n (%)	AG Thr/Ala n (%)	AA Thr/Thr n (%)	Con GG Ala/Ala n (%)	AG Thr/Ala n (%)	AA Thr/Thr n (%)
HLA DQA1*0301+	7 (23)	9 (29)	15 (48)	12 (16)	31 (43)	30 (41)
HLA DQA1*0301-	7 (16)	29 (66)	8 (18)	20 (13)	78 (50)	59 (37)
HLA DQA1*0501+	9 (17)	31 (59)	13 (25)	11 (11)	48 (47)	43 (42)
HLA DQA1*0501-	5 (23)	7 (32)	10 (46)	22 (16)	61 (44)	55 (49)
Total individuals typed for HLA DQA1	14 (19)	38 (51)	23 (31)	32 (14)	109 (47)	89 (39)

Gene frequencies in HLA DQA1*0501+ individuals:

	AD	Con
G(Ala)	49(46)[a]	70(34)
A(Thr)	57(54)	134(66)

[a]p=0.05

Phenotypic frequency HLA DQA1*0501/CTLA4 Ala[17] positivity:

	AD	Con
	40(76%)[b]	59(58%)

[b]p<0.05

allele when compared with matched controls.[21] Whereas this HLA allele does not affect the CTLA4 association in Hashimoto's thyroiditis, we found in families with type 1 diabetes, that the protective HLA DQB1*0602 allele was no longer reduced in frequency among patients with the risk enhancing CTLA4 genotype as defined by the CTLA4 microsatellite variant (see below).

CTLA4 Polymorphisms in Addison's Disease from Other European Countries

The CTLA4 association with Addison's disease has been studied by Kemp et al who observed a significant increase of the 106 bp CTLA4 microsatellite allele in only English patients with Addison's disease, whereas the comparison between Norwegian, Finnish and Estonian patients with the respective controls did not yield significant differences.[19] However, this lack of statistical significance was due to the observation, that the 106 bp CTLA4 microsatellite allele was more frequent in Norwegian (25%), Finnish (31.6%) or Estonian (35.5%) controls, in contrast to English controls (13.6%). Whereas the English control group was large (n=173), the other groups of controls were somewhat smaller (n=100/71/45). The frequency of the same allele was similar in Addison's groups: ranging from 21%

of isolated Addison's in Norway to 75% in Estonian patients with poliglandular syndrome type 2.[19] The same group found an association of the same CTLA4 microsatellite allele with vitiligo, especially when vitiligo was accompanied with other autoimmune disorders such as Addison.[20]

Thus the strength of the CTLA4 association with Addison's disease will depend on the distribution of the alleles in the background population. It is not known whether Addison's disease has a higher incidence in these Scandinavian countries, where type 1 diabetes occurs more frequently.

Findings of CTLA4 Variants in Type 1 Diabetes and Ramifications for Addison's Disease and Type 2 Diabetes

The CTLA4 microsatellite 102-106bp allele has been found to be both linked to and associated with type 1 diabetes.[22] This allele is in linkage with the exon 1 polymorphism (49 A->G, thr->ala) so that both serve as markers for the disease susceptibility. Thus the CTLA4 allele found to be associated with Addison's disease in British patients corresponds to the ala17 exon 1 variant studied by us. Investigating the transmission from parents to diabetic patients we observed a significant excess of expected transmission of the CTLA4 G49, 102 bp and promoter C-318 haplotype, indicating a susceptibility factor in German families. The combination of CTLA4 A49 and 84 bp microsatellite was less often transmitted pointing to a protective haplotype. A combined analysis of all microsatellite alleles showed them to be more informative than the exon variants. Thus the CTLA4 microsatellite allele appears to be the primary susceptibility factor. When compared with the HLA DQ alleles we found that protection by CTLA4 could not be conferred in the presence of high risk HLA DQ2 or DQ8. Thus the risk alleles of HLA DQ appear dominant over protective CTLA4 genes.

Furthermore the reduced transmission of CTLA4 A49, 84 bp microsatellite and promoter cytosine -318 was significant in mothers pointing to a parent-of-origin or imprinting effect.[22]

Recently the CTLA4 gene has also been implicated in the susceptibility to type 2 diabetes where an ala/ala 17 genotype correlated with lower body mass index.[23] Although the distribution of the CTLA4 variants was similar in patients and controls, patients with the thr/thr 17 genotype tended to develop less frequently microangiopathic lesions.[23] These findings need to be confirmed in a larger series of patients but indicate a wider application of susceptibility screening in diabetes mellitus and other endocrine disorders.

Acknowledgements

Supported by the Deutsche Forschungsgemeinschaft (DFG Ba 976/2-3 and /8-1).

References

1. Laureti S, Aubourg P, Calcinaro F et al. Etiological diagnosis of primary adrenal insufficiency using an original flowchart of immune and biochemical markers. J Clin Endocrinol Metab 1998; 83:3163-3168.
2. Laureti S, De Bellis A, Muccitelli VI et al. Levels of adrenocortical autoantibodies correlate with the degree of adrenal dysfunction in subjects with preclinical Addison's disease. J Clin Endocrinol Metab 1998; 83:3507-3511.
3. Laureti S, Vecchi L, Santeusanio F et al. Is the incidence of Addison's disease underestimated? J Clin Endocrinol Metab 1999; 84:1762
4. Kong M-F, Jeffcoate W. Comment on Is the incidence of Addison's disease underestimated? J Clin Endocrinol Metab 1999; 84:4295

5. Meyer G, Donner H, Herwig J et al. Screening for an AIRE-1 mutation in patients with Addison's disease, type 1 diabetes, Graves' disease and Hashimoto's thyroiditis as well as in APECED syndrome. Clin Endocrinol (Oxf.) 2001; 54:335-338
6. Badenhoop K, Walfish PG, Rau H et al. Susceptibility and resistance alleles of human leukocyte antigen (HLA) DQA1 and HLA DQB1 are shared in endocrine autoimmune disease. J Clin Endocrinol Metab 1995; 80:2112-2117.
7. Nistico L, Buzzetti R, Pritchard LE et al. The CTLA-4 gene region of chromosome 2q33 is linked to, and associated with, type 1 diabetes. Hum Mol Genet 1996; 5:1075-1080.
8. Reiser H, Stadecker MJ. Costimulatory B7 molecules in the pathogenesis of infectious and autoimmune diseases. N Engl J Med 1996; 335:1369-1377.
9. Gribben JG, Freeman GJ, Boussiotis VA et al. CTLA4 mediates antigen-specific apoptosis of human T cells. Proc Natl Acad Sci USA 1995; 92:811-815.
10. Karandikar NJ, Vanderlugt CL, Walunas TL et al. CTLA-4: A negative regulator of autoimmune disease. J Exp Med 1996; 184:783-788.
11. Freeman M, Weetman AP. T and B cell reactivity to adrenal antigens in autoimmune Addison's disease. Clin Exp Immunol 1992; 88:275-279.
12. Weetman AP, McGregor AM. Autoimmune thyroid disease: Further developments in our understanding. Endocr Rev 1994; 15:788-830.
13. Donner H, Rau H, Walfish PG et al. CTLA4 alanine-17 confers genetic susceptibility to Graves' disease and to type 1 diabetes mellitus. J Clin Endocrinol Metab 1997; 82:143-146.
14. Harper K, Balzano C, Rouvier E et al. CTLA-4 and CD28 activated lymphocyte molecules are closely related in both mouse and human as to sequence, message expression, gene structure, and chromosomal location. J Immunol 1991; 147:1037-1044.
15. Yanagawa T, Hidaka Y, Guimaraes V et al. CTLA-4 gene polymorphism associated with Graves' disease in a caucasian population. J Clin Endocrinol Metab 1995; 80:41-45.
16. Kotsa K, Watson P, Weetman AP. A CTLA-4 gene polymorphism is associated with both Graves' disease and autoimmune hypothyroidism. Clin Endocrinol 1997; 46:551-554.
17. Heuer M, Aust G, Ode-Hakim S et al. Different cytokine mRNA profiles in Graves' disease, Hashimoto's thyroiditis and non-autoimmune thyroid disorders determined by quantitative reverse trancriptase polymerase chain reaction (RT-PCR). Thyroid 1996; 6:97-106.
18. Matsuoka N, Eguchi K, Kawakami A et al. Lack of B7-1/BB1 and B7-2/B70 expression on thyrocytes of patients with Graves' disease. Delivery of costimulatory signals from bystander professional antigen-presenting cells. J Clin Endocrinol Metab 1996; 81:4137-4143.
19. Kemp EH, Ajjan RA, Husebye ES et al. A cytotoxic T lymphocyte antigen-4 (CTLA-4) gene polymorphism is associated with autoimmune Addison's disease in English patients. Clin Endocrinol 1998; 49:609-613.
20. Kemp EH, Ajjan RA, Waterman EA et al. Analysis of a microsatellite polymorphism of the cytotoxic T-lymphocyte antigen-4 in patients with vitiligo. Br J Dermatol 1999; 140:73-78.
21. Donner H, Braun J, Seidl C et al. Codon 17 polymorphism of the cytotoxic T lymphocyte antigen 4 gene in Hashimoto's thyroiditis and Addison's disease. J Endocrinol Metab 1997; 82:4130-4132.
22. Donner H, Seidl C, Braun J et al. CTLA4 gene haplotypes cannot protect from IDDM in the presence of high-risk HLA DQ8 or DQ2 alleles in German families. Diabetes 1998; 47:1158-1160.
23. Rau H, Braun J, Donner H et al. The codon 17 polymorphism of the CTLA4 gene in type 2 diabetes mellitus. J Clin Endocrinol Metab 2001; 86:653-655.
24. Seidl C, Donner H, Fischer B et al. CTLA4 codon 17 dimorphism in patients with rheumatoid arthritis. Tissue Antigens 1997; 51:62-66.

Chapter 9

CTLA-4 in Type 1 Diabetes Mellitus

Lorenza Nisticò, Isabella Cascino, Raffaella Buzzetti and Paolo Pozzilli

Introduction

The current etiological classification defines type 1 diabetes as a chronic hyperglycemia due to a cellular mediated immune destruction of the insulin-secreting pancreatic beta-cells. This disease is characterized by the presence at the onset of antibodies against insular molecules (islet cell antibodies (ICA), anti-insulin (IAA), anti-glutamic acid decarboxylase (GADA) and anti-tyrosine phosphatase (IA-2) and by a susceptible genotype at the HLA class II DRB1 and DQB1 genes. There is no cure and diabetic subjects require lifetime daily multiple injections of insulin to maintain glucose homeostasis.

Type 1 diabetes occurs in both sexes, but a slight male excess has been found in some populations.[1] Incidence data are available for children up to 14 years of age in 50 countries of the five continents and show a striking variation among populations and within the population of the same country. The age-adjusted incidence rates range from 0.1/100,000 per year in China and Venezuela to ~37/100,000 per year in Sardinia and Finland.[1]

It is widely accepted that type 1 diabetes results from interaction of a polygenic trait with environmental risk factors. This hypothesis has received support from different observations. Approximately one-tenth of the cases occurs in families and identical twins (that share the same genome, except for mitochondrial DNA, post-zygotic and epigenetic DNA changes) are more often concordant for the disease than dizygotic twins (that share half genome) thus suggesting the existence of inherited factors.[2-4] On the other hand, concordance is below 100% in identical twins indicating that a susceptible genome is not sufficient to develop the disease. Moreover, it is also likely that identical twins share more of their environment than non-identical twins. The contribution of genetic susceptibility has been underlined by assessment of the incidence of type 1 diabetes in children of migrants from high incidence to low incidence regions. Two studies have demonstrated that children of Sardinian heritage born and resident in Lazio and Lombardy (two Italian regions with incidence ~8/100,000 per year) have the same fourfold higher incidence as the population of origin. Moreover, children born in continental Italy of mixed couples with one Sardinian partner have an intermediate rate between those of Sardinia and of continental Italy.[5,6]

It is known that the major genes responsible for approximately 40% of genetic susceptibility are within the HLA region on chromosome 6p21 (insulin-dependent diabetes mellitus 1, IDDM1). This region includes highly polymorphic genes in tight linkage disequilibrium that encode proteins processing and presenting antigens to T lymphocytes. Initially, by case-control studies, it was shown that HLA class I alleles were increased in

CTLA-4 in Autoimmune Disease, edited by Flemming Pociot. ©2004 Eurekah.com.

type 1 diabetic subjects.[7] Subsequently, the primary role of HLA class II loci has been assessed.[8] Alleles associated with type 1 diabetes are different among populations and ethnic groups. Typically, most Caucasian type 1 diabetics have DRB1*03-DQB1*0201 and/or DRB1*04-DQB1*0302 haplotypes. Trans-racial comparison of disease associated alleles and haplotypes helps to dissect the complexity of HLA-diabetes association: a susceptible HLA genotype requires predisposing alleles at DRB1, DQB1 and DQA1 loci, there is a scale of allelic association from susceptibility through neutrality up to protection and the population frequencies of susceptible and protective haplotypes may explain the population-specific patterns of incidence and of HLA association.[9,10] Moreover, other loci within classes II and III of HLA seem to increase type 1 diabetes risk.[11-14]

The existence of non HLA-genes was suspected by genetic analysis of animal models of type 1 diabetes. The search was encouraged by technological advances, availability of a dense map of polymorphic genetic markers and of a large numbers of families with multiple affected siblings. Some whole genome linkage analyses have been conducted in the past decade that have led to the localization of several chromosome regions linked to type 1 diabetes.[15-18] Some researchers have instead adopted a more selective approach or have directly tested genes theoretically implicated in the pathogenesis of the disease. A list of loci so far mapped is reported in Table 1. In this chapter we will describe in detail the discovery of the IDDM12 (CTLA-4) locus. Results obtained in the different studies are not always overlapping and not every linkage or association has been independently replicated. The reasons for inconsistent results are discussed below. A likely explanation is the genetic heterogeneity of the disease: combinations of different genes may result in an identical phenotype. Efforts to identify the etiologic gene variations mapping in the non-HLA IDDM loci are currently ongoing.[43-45]

Exposure to environmental agents as factors initiating or precipitating beta-cell destruction has also been considered. The average daily energy intake of food items of animal origin (especially meat and diary products)[46] and the consumption in childhood[47] or by the mother during pregnancy[48] of foods with nitrosamine additives, that are toxic for beta-cells, are positively correlated with the disease incidence. Also the early introduction of cow's milk, due to short duration of breast feeding, increases the risk for childhood type 1 diabetes.[49] The underlying hypothesis is that the bovine beta-casein, whose sequence differs from the human protein, may pass through the immature mucosa of the gut and trigger a cellular and humoral immune response cross-reacting with a beta-cell antigen.[50]

Viral infections (rubella, coxsackie, and cytomegalovirus), contracted during fetal life, have been associated with increased risk of type 1 diabetes.[51-53] The discovery of similarities in aminoacid sequences of viral proteins and beta-cell components (insulin, glutamic acid decarboxilase and other unidentified proteins) has supported the hypothesis of molecular mimicry.[54] A mechanism of direct cytotoxicity may also involved.

The Discovery of the CTLA-4 Association with Type 1 Diabetes

It all began in the early '90s when we joined Roberto Tosi at the CNR in Rome who was planning to perform an identical-by-descent linkage analysis of type 1 diabetes in Italian affected sib-pair families. We chose to study a dozen chromosome regions that either contained genes involved in autoimmune response or were orthologous to murine or rat genome regions where susceptibility loci to autoimmune diabetes had been mapped. One region was the long arm of human chromosome 2 that corresponds to murine chromosome 1, where a locus for periinsulitis in non-obese diabetic (NOD) mouse, named Idd5, had been reported.[55,56]

Table 1. Loci and genes linked to or associated with type 1 diabetes

Locus/Gene	Chromosome	References
IDDM1	6p21	7-10
IDDM2	11p15	19-22
IDDM3	15q26	23-24
IDDM4	11q13	15-16, 24
IDDM5	6q23-q24	15, 24-26
IDDM6	18q21	15, 27-28
IDDM7	2q31	24, 29-31
IDDM8	6q27	15, 24-26, 32
IDDM9	3q22-q25	18
IDDM10	10p13-q11	18, 33
IDDM11	14q24.3-q31	34
IDDM12	2q33	see Tables 2 and 3
IDDM13	2q34	31, 35
IDDM15	6q21	26
IDDM17	10q25	36
IDDM18	5q33-34	37
Glucokinase	7	38
Vitamin D receptor	12q14	39-41
IDDMX	Xp22-p11	42
	1q	17
	16q22-q24	18

For IDDM14 and IDDM16 no public information is available

Initially, we tested microsatellite markers mapping in 2q12-q22 and in 2q33-35. The latter region gave evidence of increased allele sharing in our Italian sib-pair families and we presented these preliminary results in scientific meetings from 1994.[57-60] Detection of linkage indicates that a disease gene should lie in a rather extended region (several centiMorgans) around the positive markers. We were lucky because microsatellites linked in our data set included an (AT) repeat in the 3' untranslated (UTR) sequence of the CTLA-4 gene. CTLA-4 had been identified in 1987 by Brunet et al[61] as a transcript of cytotoxic T cells and at the time of our observation its role in regulating T cell activation was still controversial. It was also known that CD28, another T cell co-stimulatory molecule structurally homologous to CTLA-4, mapped close to CTLA-4 on chromosome 2.[62,63] Thus, it was clear to us that CTLA-4 and CD28 deserved to be considered type 1 diabetes candidate genes.

Following a suggestion by John A. Todd, we tested the CTLA-4 microsatellite for association in the presence of linkage with the disease by the transmission disequilibrium test (TDT). The second most frequent allele, that we named 104 mobility units (mu), showed increased transmission, albeit not significant, to diabetic offspring in Italian multiplex families (20 transmissions versus 10 non transmissions, $p=0.07$). We felt encouraged and undertook the collection of Italian "simplex" families (one diabetic child, his/her parents and a non-affected child, if available), a more abundant source suitable for TDT.

Searching the Genbank we found that a CTLA-4 sequence submission (accession number L15006) reported three "conflicts", i.e., potential single base polymorphisms, at positions 49, 272 and 439. Only the A49G variation, that codes for a threonine to alanine change in the signal peptide, turned out to be a true allelic difference (frequency of G allele in Italian population is approximately 30%). Preliminary analysis on 104 Italian type 1 diabetes families showed that the G49 allele, that is in tight linkage disequilibrium (LD) with the 104 mu allele, was significantly more transmitted to affected children ($p<0.025$).[64] We continued collecting and typing Italian families and established collaboration with foreign groups: 44 Spanish type 1 diabetic families were tested for the A49G single nucleotide polymorphism (SNP). The results (40 transmissions versus 18 non-transmissions to all affected siblings, $p=0.004$, 21 transmissions and 25 non-transmissions to non-affected children) replicated the data obtained in Italian families. In the meantime, an article came out that showed a significant increase of the second most frequent allele of the CTLA-4 microsatellite (the Authors named this allele 106 base-pairs (bp)) in Caucasian patients affected by Graves' disease.[65] This independent observation strengthened the hypothesis that a gene controlling a step in an autoimmune process was in LD with CTLA-4, if not CTLA-4 itself.

In collaboration with Todd's group, the A49G SNP was typed in 284 UK, 180 US and 123 Sardinian type 1 diabetic families and in a panel of Chinese patients affected by Graves' disease and controls. TDT of G49 allele in diabetic families was not statistically significant, although a trend toward increased transmissions was observed in the US data set (177 transmissions versus 145 non-transmissions, $p=0.074$). Instead, the Chinese case-control study was significant and in agreement with Italian and Spanish data. We typed 83 additional Italian families and TDT of the G49 allele in the whole Italian family panel ($n=187$) was significant (114 transmissions and 75 non-transmissions to all affected children, $p=0.004$). To obtain a valid test of association, we analyzed transmissions to Italian and Spanish probands only (one affected child in multiplex families) and obtained significant data (112 transmissions versus 65 non-transmissions, $p=0.0004$). Importantly, no transmission distortion was observed to non-diabetic sibs. The overall TDT to affected siblings of the five family data sets was significant ($p=0.002$).

The Belgian Diabetes Registry, which is a large population-based collection of type 1 diabetic patients, was also analyzed. Belgian patients ($n=483$) had significantly higher frequencies of G allele, G phenotype and GG genotype compared to controls ($n=529$). The overall data (three out of six diabetic panels studied were significant and consistent) provided evidence for an association of the CTLA-4 region with type 1 diabetes and the locus was named IDDM12. Results of this collaborative effort were joined and published in 1996.[66]

Later on the UK, US, Italian and Sardinian family collections, previously typed for the A49G variation, plus 185 additional families, were also analyzed for the CTLA-4 microsatellite. The two most frequent alleles showed percentage of transmission significantly different from random expectation: alleles 262 and 280 (the latter corresponds to allele 104 mu or 106 bp) showed decreased and increased transmissions, respectively, to all diabetic children either in the overall data set (allele 262, $p=0.009$; allele 280, $p=0.001$) or in the UK and US families combined (allele 262, $p=0.001$; allele 280, $p=0.02$).[31] These data indicate that the CTLA-4 microsatellite is a better disease risk marker than the A49G variation.

Genetic Studies on CTLA-4 and Human Type 1 Diabetes

Since our first observations, many papers testing the hypothesis of CTLA-4 association with different autoimmune diseases in diverse populations followed one another.[67] To our knowledge, sixteen articles that studied the association with type 1 diabetes have been published up to the year 2000 (Tables 1 and 2). Alternate conclusions have been drawn. There are some explanations for inconsistent results among distinct populations and among different data sets from the same population.

First, the study should have enough power (that is probability) to detect association at a statistically significant level. Power depends on the relative risk and on the frequency in the general population of the factor tested and on the size of the sample studied (either families or case-controls). Type 1 diabetes genes, apart from HLA, are estimated to confer low risk (i.e., 1.5), thus for an allele frequency ranging between 10% to 50%, 700 to 300 patients and controls should be analyzed (for an 80% power and a significance level of 0.05). Studying data set of inappropriate size reduces the chance of detecting true small effects (false negative).

There may be also population specific factors such as pattern and level of LD between the marker allele tested and the disease-predisposing allele. Population isolates have a higher degree of LD than cosmopolitan populations.[83]

Finally, locus heterogeneity must be considered: a gene may account only for a fraction of the cases depending on genetic background and on environmental factors. In this respect, some researchers studied CTLA-4 association sub-grouping diabetic patients on the basis of genetic, immune and clinical features. Some conflicting results have been reported but these are probably due to the small number of cases included in the sub-classes and to non-stringent statistic thresholds (p values not corrected for number of comparisons performed). In the largest study of this type the Authors found no difference in the CTLA-4 G49 allele distribution with respect to age of disease onset, HLA-DR, -DQ and insulin genotypes and the presence of islet or thyrogastric autoantibodies.[73]

Role of CTLA-4 in Human Type 1 Diabetes

There is no direct proof that CTLA-4 is a diabetes susceptibility gene. CTLA-4 gene itself may be primarily involved in the disease pathogenesis if one or more of its polymorphisms determine an effect on the protein function that may be relevant to the disease process. Alternatively, it is possible that CTLA-4 is in LD with the etiologic variation residing somewhere in the region.

Genetic studies on CTLA-4 have focused on three polymorphisms, so far: a C-T SNP in position -318 from the ATG start codon[84], the A49G in exon 1 and the dinucleotide repeat in the 3'-UTR. Recently, Marron et al[69] described another SNP in CTLA-4 intron 1 (C/T in position -819 from exon 2-start site) that showed more significant association with type 1 diabetes than the A49G and the microsatellite.

The biological significance of these polymorphisms is not known. The C-318T variation does not affect any known consensus sequence in the regulatory region of the promoter.[84]

As far as the A49G variation is concerned, Kouki et al[85] demonstrated that proliferative response of T lymphocytes is increased in the presence of soluble blocking anti-human CTLA-4 monoclonal antibody. The augmentation is significantly higher in individuals A/A homozygous compared with G/G subjects regardless of the affection status. This is the first report that shows that CTLA-4 effects can be modulated by its genotype, however the phenotype observed cannot be directly related to exon 1 variation due to its strong LD with the (AT)n polymorphism at 3'-UTR, the C/T SNP described by Marron et al[69] and, possibly, with others yet unknown.

Table 2. CTLA-4 association with type 1 diabetes: family studies

Population	Families	Polymorphisms typed																References
		promoter -318T allele			exon 1 49G allele				intron 1 T allele				3' UTR (AT)n					
		TDT			TDT			Tsp	TDT			Tsp		TDT			Tsp	
	N	T	NT	P value	T	NT	P value	P value	T	NT	P value	P value	alleles	T	NT	P value	P value	
Italian	187				114	75	0.004											66
Spanish	44				40	18	0.004											
UK	284				264	253	NS											
US	180				177	145	NS											
Sardinian	123				46	45	NS											
TOTAL	818				641	536	0.002											
US Caucasian[a]	295				297	257	NS	NS	307	249	<0.05	0.02	8[b]	230	305	<0.005	0.002	68, 69
European+Mexican American	164				103	70	<0.05	0.02	94	64	<0.05	0.02	8[b]	82	106	NS	0.02	
Asian	69				18	16	NS	NS	19	16	NS	NS	8[b]	16	13	NS	NS	
TOTAL	528				418	343	<0.05	0.01	420	329	<0.005	0.0009	8[b]	328	424	<0.005	0.03	
German	109	21	19	NS	71	39	<0.05						84[b]	42	75	<0.05		70
													102[c]	55	33	NS		
UK+US[a]	588												262[b]			0.001		31
													280[c]			0.02		
UK+US+Italian+Sardinian[a]	960												262[b]			0.009	0.004	
													280[c]			0.001	0.004	
Basque	63				30	24	NS						262[b]	22	36	NS		71
													280[c]	17	24	NS		
Danish	254				166	145	NS						148[c]	91	83	NS		72
Spanish	39				17	26	NS											

TDT: Transmission disequilibrium test. Tsp is a modified TDT and in multiplex families only TSP is a valid test of association. T and NT: Transmissions and non transmissions from heterozygous parents for the indicated allele to the affected children. a: Partial overlap with datasets in ref. 66. b: 8, 84 and 262 indicate the same allele. c: 102, 280 and 148 indicate the same allele. NS: not significant (P>0.05).

With regard to the microsatellite, it has been speculated that (AT)n length in 3'-UTR can affect mRNA stability, but this has not been proven for CTLA-4.

We and others[84,86] have searched CTLA-4 exons and part of introns for additional polymorphisms: nothing has been found in coding regions. We have identified two new C-T SNPs in 5' sequence and in intron 1, respectively, but they did not show significant transmission distortion in 200 Italian diabetic families (unpublished).

Sequencing of cDNA from human and rodents T and B lymphocytes revealed a CTLA-4 alternate transcript that lacks the transmembrane domain[87,88] and results in a soluble form (sCTLA-4) detectable in the sera. Oaks et al[89] have also demonstrated that serum levels of sCTLA-4 are higher in patients with autoimmune thyroid diseases compared to healthy subjects. It is not clear whether this phenomenon is due to DNA differences or to an indirect effect related to the affection status.

Finally, it cannot be ruled out that additional polymorphisms map in flanking and intron sequences of CTLA-4 locus and that these, taken singly or in combination, could be the etiologic variation.

Genetic Studies in Animals

A great deal of information on CTLA-4 function and on its possible role in the pathogenesis of autoimmune diabetes comes from genetic and functional studies in mice.

Animal models of human disorders constitute a powerful tool for investigation of the mechanisms involved in complex diseases since confounders acting in research on human populations are not present in strains. Being inbred, the genetic background is homogeneous and fixed among generations and sibship. Also the environment, which probably plays an important role in the unfolding of human diseases, is controlled and homogeneous for laboratory animals. Moreover, information can be drawn by manipulating the genetic background of the animal and producing extreme phenotypic consequences.

Non-obese diabetic (NOD) mouse is a strain that spontaneously develops autoimmune diabetes with many similarities to the human disorder. The Ctla-4 gene possibly controls insulitis and diabetes in NOD mouse since it maps within the Idd5 diabetes susceptibility locus. Idd5 had been identified in genomic screens conducted on NOD mouse as a large (34 cM) region on proximal chromosome 1 (orthologous to human chromosome 2q) that was linked to insulitis.[55,56] To identify the minimal region containing the susceptibility gene, Hill et al[90] have recently developed some congenic strains by backcrossing NOD mice (diabetes prone) to a B10 (diabetes resistant) strain and selecting the progeny that has NOD genetic background except for the region of interest on chromosome 1 that is derived from the B10 resistant animals. NOD.B10 Idd5 mice have a lower frequency of diabetes than NOD animals. Additional congenic strains with an even shorter segment of the B10-derived chromosome 1 have then been generated. Comparing diabetes incidence among different strains it was possible to narrow the region(s) containing the gene(s) of resistance to the disease. They have determined that Idd5 is actually a two-gene locus. Idd5.1 is a 1.5 cM region that includes four genes: Casp8, Cflar, Cd28 and Ctla-4 and corresponds to IDDM12 region identified in humans. Idd5.2 contains Nramp1 and Il-8ra (interleukin-8 receptor a). No coding variations have been found between NOD and B10 sequences of Cflar, Cd28 and Ctla-4. A conservative Ala-Val substitution has been found in Casp8, but its role in disease has not been explored. Thus, similarly to what has been seen in humans, it is possible that the etiologic variation resides in non-translated sequences around these genes.

Table 3. CTLA-4 association with type 1 diabetes: case-control studies

Population	Cases	Controls	Polymorphisms typed					References
			exon 1 A49G			3' UTR (AT)n		
	N	N	Genotypes*	G positives*	G allele*	Genotypes*	Alleles*	
Belgian	483	529	0.0003	0.0002	<0.0001			66, 73
Caucasian	293	325	0.01	0.009	0.004			74
Florida US	244	274	NS		NS			68
Spanish	89	57	NS		NS			
Korean	97	112	NS		NS			
Chinese	180	379	NS		NS			
Caucasian American[a]	160	80					NS[b]	75
Polish	192	136	0.002	0.002				76
Japanese	173	425	NS		NS			77
Japanese	111	145	NS		NS			78
Japanese	117	141	NS		NS			79
Japanese	110	200	NS	NS	NS			80
Swedish	616	502				0.0008[c]	0.0009[c]	81
Chinese	253	91	0.009	0.002	0.005			82

* Values in these columns are statistical significances (P values); a: HLA selected; b: for allele 106 bp; c: these authors compare groups of "short", "intermediate" and "very long" alleles; NS: not significant (P>0.05).

Functional Studies in Mice

The controversy concerning the role of Ctla-4 in the regulation of T cell activation has been resolved by generating Ctla-4 knock-out mice, that are mice in which both copies of the Ctla-4 genes have been disrupted by homologous recombination. These animals have an extensive accumulation of activated lymphocytes in several tissues and organs. Infiltrating cells express activation markers and exhibit high proliferation rates in response to stimuli or spontaneously thus implying that Ctla-4 acts as a negative regulator of T cells.[91]

The NOD model of autoimmune diabetes has been exploited to understand how and when Ctla-4 exert its immunoregulatory role.

Luhder et al[92] demonstrated in a TCR transgenic model of autoimmune diabetes that blockade with anti-Ctla-4 monoclonal antibody (mAb) accelerated the development of diabetes. They also established that Ctla-4 does not act on naive cells, but in a narrow time window, when activated T cells migrate to the pancreatic islets and re-encounter the antigen for a second time.[93] This is consistent with the observation that maximal Ctla-4 protein expression is seen 48-72 hours later T cell activation. A sub-population of $CD4^+$ T lymphocytes, that is $CD25^+$ and constitutively expresses Ctla-4, probably exerts the inhibitory role. This T cell subset is reduced in NOD mice compared to other mouse strains. Moreover, if NOD splenocytes depleted of $CD4^+$ $CD25^+$ Ctla-4^+ T cells are transferred into NOD.SCID mice an increase in the frequency of diabetes is observed compared to animals that received total splenocytes.[94] Interestingly, presence of $CD4^+$ $CD25^+$ Ctla-4^+ cells requires the integrity of the CD28/B7 system. Regulatory T cells positive for CD25, CD4 and Ctla-4 suppressively control other T cells by two possible mechanisms: they may compete, through Ctla-4, with other T cells for the costimulatory signal on antigen presenting cells. Alternatively, regulatory T cells deliver to other T cells a negative signal of proliferation or activation.[95]

Conclusion

Much detailed information has been gained in the last few years on CTLA-4, and its function is compatible with a role in the pathogenesis of type 1 diabetes and other immune-mediated diseases. However, genetic data neither show nor rule out in a definite and conclusive manner that CTLA-4 and/or the flanking genes CD28 and ICOS are primarily responsible for the observed disease association. Sequencing of the murine and human CTLA-4 genes in a diabetes prone strain and in affected people has demonstrated that the extracellular binding domain and the cytoplasmic tail that delivers the signal are not modified in this disease, thus the explanation must be searched for elsewhere. Possibly, every SNP in the region has to be looked for and genetically tested. In tight conjunction, studies of functional genomics are needed to investigate the effect encoded by the sequence variations of the CTLA-4 locus that are found associated with the disease. Slight modifications of the transcribed or translated product are possibly involved and such evidence may be difficult to detect and discriminate from secondary effects. Thus very sensitive methods and artificial systems may be required to demonstrate a direct relationship between the genotype and the biological effect at the expression/protein level. Correlation of the protein phenotype with the disease outcome is a further step that should take into account the contribution of the other disease genes and of the environmental factors.

Acknowledgements

We are deeply grateful to Roberto Tosi. A special thank goes to John A. Todd, Lynn Pritchard and Francesco Cucca. We wish also to acknowledge the physicians of the IMDIAB group, Maria Teresa Martinez, Manuel Serrano-Rios, Frans Gorus, Bart Van der Auwera and Emanuele Bosi for sharing their samples and results and Richard H. Butler for revising this manuscript.

References

1. Karvonen M, Viik-Kajander M, Moltchanova E et al. Incidence of childhood type 1 diabetes worldwide. Diabetes care 2000; 23:1516-1526.
2. Barnett AH, Eff C, Leslie RDG et al. Diabetes in identical twins: a study of 200 pairs. Diabetologia 1981; 20:87-93.
3. Olmos P, A'Hern R, Heaton DA et al. The significance of the concordance rate for type 1 (insulin-dependent) diabetes in identical twins. Diabetologia 1988; 31:747-50.

4. Kyvik KO, Green A, Beck-Nielsen H. Concordance rates of insulin dependent diabetes mellitus: a population based study of young Danish twins. BMJ 1995; 311:913-917.
5. Muntoni SA, Fonte MT, Stoduto S et al. Incidence of insulin dependent type 1 diabetes mellitus among Sardinian heritage children born in Lazio region (Italy). Lancet 1997; 349:160-162.
6. Calori G, Gallus G, Bognetti E et al. Insulin-dependent diabetes mellitus in Sardinian heritage children living in Lombardy. Lancet 1998; 351:263-264.
7. Singal DP, Blajchman MA. Histocompatibility (HL-A) antigens, lymphocytotoxic antibodies and tissue antibodies in patients with diabetes mellitus. Diabetes 1973; 22: 429-432.
8. Nerup J, Platz P, Andersen OO et al. HL-A antigens and diabetes mellitus. Lancet 1974; 2: 864-866.
9. Cucca F, Todd JA. HLA susceptibility to type 1 diabetes: methods and mechanisms. In: Browning M, McMichael A, eds. HLA and MHC: genes, molecules and function. Oxford: Bios Scientific Publishers Ltd, 1996.
10. Undlien DE, Lie BA, Thorsby E. HLA complex genes in type 1 diabetes and other autoimmune diseases. Which genes are involved? Trends Genet 2001; 17:93-100.
11. Erlich HA, Rotter JI, Chang JD et al. Association of HLA-DPB1*0301 with IDDM in Mexican-Americans. Diabetes 1996; 45:610-614.
12. Noble JA, Valdes AM, Cook M et al. The role of HLA class II genes in insulin-dependent diabetes mellitus: molecular analysis of 180 Caucasian, multiplex families. Am J Hum Genet 1996; 59:1134-1148.
13. Cucca F, Dudbridge F, Loddo M et al. The HLA-DPB1-associated component of the IDDM1 and its relationship to the major loci HLA-DQB1, -DQA1, and -DRB1. Diabetes 2001; 50:1200-1205.
14. Zavattari P, Lampis R, Motzo C et al. Conditional linkage disequilibrium analysis of a complex disease superlocus, IDDM1 in the HLA region, reveals the presence of independent modifying gene effects influencing the type 1 diabetes risk encoded by the major HLA-DQB1, -DRB1 disease loci. Hum Mol Genet 2001; 10:881-889.
15. Davies JL, Kawaguchi Y, Bennett ST et al. A genome-wide search for human type 1 diabetes susceptibility genes. Nature 1994; 371:130-136.
16. Hashimoto L, Habita C, Beressi JP et al. Genetic mapping of a susceptibility locus for insulin-dependent diabetes mellitus on chromosome 11q. Nature 1994; 371:161-164.
17. Concannon P, Gogolin-Ewens KJ, Hinds DA et al. A second-generation screen of the human genome for susceptibility to insulin-dependent diabetes mellitus. Nat Genet 1998; 19:292-296.
18. Mein CA, Esposito L, Dunn MG et al. A search for type 1 diabetes susceptibility genes in families from the United Kingdom. Nat Genet 1998; 19:297-300.
19. Bell GI, Horita S, Karam JH. A polymorphic locus near the human insulin gene is associated with insulin-dependent diabetes mellitus. Diabetes 1984; 33:176-183.
20. Catignani Kennedy G, German MS, Rutter WJ. The minisatellite in the diabetes susceptibility locus IDDM2 regulates insulin transcription. Nat Genet 1995; 9:293-298.
21. Pugliese A, Zeller M, Fernandez A Jr et al. The insulin gene is transcribed in the human thymus and transcription levels correlated with allelic variation at the INS VNTR-IDDM2 susceptibility locus for type 1 diabetes. Nat Genet 1997; 15:293-297.
22. Vafiadis P, Bennett ST, Todd JA et al. Insulin expression in human thymus is modulated by INS VNTR alleles at the IDDM2 locus. Nat Genet 1997; 15:289-292.
23. Field LL, Tobias R, Magnus T. A locus on chromosome 15q26 (IDDM3) produces susceptibility to insulin-dependent diabetes mellitus. Nat Genet 1994; 8:189-194.
24. Luo DF, Buzzetti R, Rotter J et al. Confirmation of three susceptibility genes to insulin-dependent diabetes millitus: IDDM4, IDDM5 and IDDM8. Hum Mol Genet 1996; 5:693-698.
25. Davies JL, Cucca F, Goy JV et al. Saturation multipoint linkage mapping of chromosome 6q in type 1 diabetes. Hum Molec Genet 1996; 5:1071-1074.
26. Delepine M, Pociot F, Habita C et al. Evidence of a non-MHC susceptibility locus il type 1 diabetes linked to HLA on chromosome 6. Am J Hum Genet 1997; 60:174-187.
27. Merriman T, Twells R, Merriman Met al. Evidence by allelic association-dependent methods for a type 1 diabetes polygene (IDDM6) on chromosome 18q21. Hum Mol Genet 1997; 6:1003-1010.
28. Merriman TR, Eaves IA, Twells RC et al. Transmission of haplotypes of microsatellite markers rather than single marker alleles in the mapping of a putative type 1 diabetes susceptibility gene (IDDM6). Hum Mol Genet 1998; 7:517-524.
29. Copeman JB, Cucca F, Hearne CM et al. Linkage disequilibrium mapping of a type 1 diabetes susceptibility gene (IDDM7) to chromosome 2q31-33. Nat Genet 1995; 9:80-85.

30. Luo DF, Maclaren NK, Huang HS et al. Intrafamilial and case-control association analysis of D2S152 in insulin-dependent diabetes. Autoimmunity 1995; 21:143-147.
31. Esposito L, Hill NJ, Pritchard LE et al. Genetic analysis of chromosome 2 in type 1 diabetes. Analysis of putative loci IDDM7, IDDM12, and IDDM13 and candidate genes NRAMP1 and IA-2 and the Interleukin-1 gene cluster. Diabetes 1998; 47:1797-1799.
32. Luo DF, Bui MM, Muir A et al. Affected-sib-pair mapping of a novel susceptibility gene to insulin-dependent diabetes mellitus (IDDM8) on chromosome 6q25-q27. Am J Hum Genet 1995; 57:911-919.
33. Reed P, Cucca F, Jenkins Set al. Evidence for a type 1 diabetes susceptibility locus (IDDM10) on human chromosome 10p11-q11. Hum Mol Genet 1997; 6:1011-1016.
34. Field LL, Tobias R, Thomson G et al. Susceptibility to insulin-dependent diabetes mellitus maps to a locus (IDDM11) on human chromosome 14q24.3-q31. Genomics 1996; 33:1-8.
35. Morahan G, Huang D, Tait BD et al. Markers on distal chromosome 2q linked to insulin-dependent diabetes mellitus. Science 1996; 272:1811-1813.
36. Verge CF, Vardi P, Babu S et al. Evidence for oligogenic inheritance of type 1 diabetes in a large Bedouin Arab family. J Clin Invest 1998; 102:1569-1575.
37. Morahan G, Huang D, Ymer SI et al. Linkage disequilibrium of a type 1 diabetes susceptibility locus with a regulatory IL12B allele. Nat Genet 2001; 27:218-221.
38. Rowe RE, Wapelhorst B, Bell GI et al. Linkage and association between insulin-dependent diabetes mellitus (IDDM) susceptibility and markers near the glucokinase gene on chromosome 7. Nat Genet 1995; 10: 240-245.
39. McDermott MF, Ramachandran A, Ogunkolade BW et al. Allelic variation in the vitamin D receptor influences susceptibility to IDDM in Indian Asians. Diabetologia 1997; 40:971-975.
40. Pani Ma, Knapp M, Donner H et al. Vitamin D receptor allele combinations influence genetic susceptibility to type 1 diabetes in Germans. Diabetes 2000; 49:504-507.
41. Chang TJ, Lei HH, Yeh JI et al. Vitamin D receptor gene polymorphisms influence susceptibility to type 1 diabetes mellitus in the Taiwanese population. Clin Endocrinol 2000; 52:575-580.
42. Cucca F, Goy JV, Kawaguchi Y et al. A male-female bias in type 1 diabetes and linkage to chromosome Xp in MHC HLA-DR3-positive patients. Nat Genet 1998; 19:301-302.
43. Nakagawa Y, Kawaguchi Y, Twells RCJ et al. Fine mapping of the diabetes-susceptibility locus, IDDM4, on chromosome 11q13. Am J Hum Genet 1998; 63:547-556.
44. Eckenrode S, Marron MP, Nicholls R et al. Fine-mapping of the type 1 diabetes locus (IDDM4) on chromosome11q and evaluation of two candidate genes (FADD and GALN) by affected sibpair and linkage-disequilibrium analyses. Hum Genet 2000; 106:14-18.
45. Owerbach D. Physical and genetic mapping of IDDM8 on chromosome 6q27. Diabetes 2000; 49:508-512.
46. Muntoni SA, Cocco P, Aru G et al. Nutritional factors and worldwide incidence of childhood type 1 diabetes. Am J Clin Nutr 2000; 71:1525-1529.
47. Dahlquist GG, Blom LG, Persson LA et al. Dietary factors and the risk of developing insulin dependent diabetes in childhood. BMJ 1990; 300:1302-1306.
48. Helgason T, Jonasson MR. Evidence for a food additive as a cause of ketosis-prone diabetes. Lancet 1981; 2:716-20.
49. Borch-Johnsen K, Joner G, Mandrup-Poulsen T et al. Relation between breast-feeding and incidence rates of insulin-dependent diabetes mellitus. A hypothesis. Lancet 1984; 2:1083-6.
50. Cavallo MG, Fava D, Monetini L et al. Cell-mediated immune response to beta casein in recent-onset insulin-dependent diabetes: implications for disease pathogenesis. Lancet 1996; 348:926-928.
51. Menser MA, Forrest JM, Bransby RD. Rubella infection and diabetes mellitus. Lancet 1978; 1:57-60.
52. Dahlquist GG, Ivarsson S, Lindberg B et al. Maternal enteroviral infection during pregnancy as a risk factor for childhood IDDM. A population-based case-control study. Diabetes 1995; 44:408-413.
53. Pak CY, Eun HM, McArthur RG et al. Association of cytomegalovirus infection with autoimmune type 1 diabetes. Lancet 1988; 2:1-4.
54. Kukreja A, Maclaren NK. Current cases in which epitope mimicry is considered as a component cause of autoimmune disease: immune-mediated (type 1) diabetes. Cell Mol Life Sci 2000; 57:534-541.
55. Garchon H-J, Bedossa P, Eloy L et al. Identification and mapping to chromosome 1 of a susceptibility locus for periinsulitis in non-obese diabetic mice. Nature 1991; 353:260-262.

56. Cornall RJ, Prins J-B, Todd JA et al. Type 1 diabetes in mice is linked to the interleukin-1 receptor and Lsh/Ity/Bcg genes on chromosome 1. Nature 1991; 353:262-265.
57. Nisticò L, Buzzetti R, Latini L et al. Linkage analysis in IDDM affected sib-pair from Continental Italy. In: Abstract book of the 13th International Immunology and Diabetes Workshop. 1994:82.
58. Buzzetti R, Nisticò L, Pozzilli P et al. CTLA-4 microsatellite identifies a new region on chromosome 2 linked to IDDM. Diabetologia 1995; 38:A29.
59. Nisticò L, Buzzetti R, Pozzilli P et al. Identification of a new region linked to IDDM in affected multiplex families from the Mediterranean area. Diabetes 1995; 44 (suppl.1):17A (abstract 55).
60. Buzzetti R, Nisticò L, Mesturino C et al. A novel region of susceptibility to IDDM on chromosome 2. J Endocrinol Invest 1995; 18(suppl. to no. 5): 3 (abstract 3).
61. Brunet J-F, Denizot F, Luciani M-F et al. A new member of the immunoglobulin superfamily-CTLA-4. Nature 1987; 328:267-270.
62. Harper K, Balzano C, Rouvier E et al. CTLA-4 and CD28 activated lymphocyte molecules are closely related in both mouse and human as to sequence, message expression, gene structure, and chromosomal location. J Immunol 1991; 147:1037-1044.
63. Buonavista N, Balzano C, Pontarotti P et al. Molecular linkage of the human CTLA4 and CD28 Ig-superfamily genes in yeast artificial chromosomes. Genomics 1992; 13:856-861.
64. Nisticò L, Buzzetti R, Mesturino C et al. Linkage and association of CTLA-4 gene to IDDM in the Italian population. Autoimmunity 1995; 21(1):18 (abstract 070).
65. Yanagawa T, Hidaka Y, Guimaraes V et al. CTLA-4 gene polymorphism associated with Graves' disease in a Caucasian population. J Clin Endocrinol Metabol 1995; 80:41-45.
66. Nisticò L, Buzzetti R, Pritchard LA et al. The CTLA-4 gene region of chromosome 2q33 is linked to, and associated with type 1 diabetes. Hum Molec Genet 1996; 5:1075-1080.
67. Kristiansen OP, Larsen ZM, Pociot F. CTLA-4 in autoimmune diseases—a general susceptibility gene to autoimmunity? Genes Immunity 2000; 1:170-184.
68. Marron MP, Raffel LJ, Garchon HJ et al. Insulin-dependent diabetes mellitus (IDDM) is associated with CTLA4 polymorphisms in multiple ethnic groups. Hum Molec Genet 1997; 6:1275-1282.
69. Marron MP, Zeidler A, Raffel LJ et al. Genetic and physical mapping of a type 1 diabetes susceptibility gene (IDDM12) to a 100kb PAC clone containing D2S72-CTLA4-D2S105 on chromosome 2q33. Diabetes 2000; 49:492-499.
70. Donner H, Seidl C, Braun J et al. CTLA4 gene haplotypes cannot protect from IDDM in the presence of high risk HLA DQ8 or DQ2 alleles in German families. Diabetes 1998; 47: 1158-1160.
71. Perez de Nanclarez G, Bilbao JR, Nisticò L et al. No evidence of association of chromosome 2q with type 1 diabetes in the Basque population. Diabetologia 1999; 42:119-120.
72. Larsen ZM, Kristiansen OP, Mato E et al. IDDM12 (CTLA4) on 2q33 and IDDM13 on 2q34 in genetic susceptibility to type 1 diabetes (insulin dependent). Autoimmunity 1999; 31:35-42.
73. Van der Auwera BJ, Vandewalle CL, Schuit FC et al. CTLA-4 gene polymorphism confers susceptibility to insulin dependent diabetes mellitus (IDDM) independently from age and from other genetic or immune disease markers. Clin Exp Immunol 1997; 110:98-103.
74. Donner H, Rau H, Walfish PG et al. CTLA4 alanine -17 confers susceptibility to Graves' disease and to type 1 diabetes. J Clin Endocrinol Metab 1997; 82:143-146.
75. Owerbach D, Naya FJ, Tsai M-J et al. Analysis of candidate genes for susceptibility to type 1 diabetes. A case-control and family association study of genes on chromosome 2q31-35. Diabetes 1997; 46:1069-1074.
76. Krokowski M, Bodalski J, Machejko P et al. CTLA-4 gene polymorphism is associated with predisposition to IDDM in a population from central Poland. Diabetes Metab 1998; 24:241-243.
77. Awata T, Kurihara S, Iitaka M et al. Association of CTLA-4 gene A-G polymorphism (IDDM12 locus) with acute-onset and insulin-depleted IDDM as well as autoimmune thyroid disease (Graves' disease and Hashimoto's thyroiditis) in the Japanese population. Diabetes 1997; 47:128-129.
78. Abe T, Takino H, Yamasaki H et al. CTLA-4 gene polymorphism correlates with the mode of onset and presence of ICA512 Ab in Japanese type 1 diabetes. Diabetes Res Clin Pract 1999; 46:169-175.
79. Hayashi H, Kusaka I, Nagasaka S et al. Association of CTLA-4 polymorphism with positive anti-GAD antibody in Japanese subjects with type 1 diabetes mellitus. Clin Endocrinol 1999; 51:793-799.
80. Yanagawa T, Maruyama T, Gomi K et al. Lack of association between CTLA-4 gene polymorphism and IDDM in Japanese subjects. Autoimmunity 1999; 29:53-56.

81. Lowe RM, Graham J, Sund G et al. The length of the CTLA-4 microsatellite (AT)n-repeat affects the risk for type 1 diabetes. Autoimmunity 2000; 32:173-180.
82. Lee Y-J, Huang F-Y, Lo F-S et al. Association of CTLA-4 gene A-G polymorphism with type 1 diabetes in Chinese children. Clin Endocrinol 2000; 52:153-157.
83. Zavattari P, Deidda E, Whalen M et al. Major factor influencing linkage disequilibrium by analysis of different chromosome regions in distinct populations: demography, chromosome recombination frequency and selection. Hum Molec Genet 2000; 9:2947-2957.
84. Deichman K, Heinzmann A, Bruggenolte E et al. An Mse I RFLP in the human CTLA4 promoter. Biochem Biophys Res Commun 1996; 225:817-818.
85. Kouki T, Sawai Y, Gardine C et al. CTLA-4 gene polymorphism at position 49 in exon 1 reduces the inhibitory function of CTLA-4 and contributes to the pathogenesis of Graves' disease. J Immunol 2000; 165:6606-6611.
86. Ligers A, Xu C, Saarinen S et al. The CTLA-4 gene is associated with multiple sclerosis. J Neuroimmunol 1999; 97:182-190.
87. Magistrelli G, Jeannin P, Herbault N et al. A soluble form of CTLA-4 generated by alternative splicing is expressed by nonstimulated human T cells. Eur J Immunol 1999; 29:3596-3602.
88. Oaks M, Hallet KM, Todd Penwell R et al. A native soluble form of CTLA-4. Cell Immunol 2000; 201:144-153.
89. Oaks MK et Hallet KM. Cutting edge: a soluble form of CTLA-4 in patients with autoimmune thyroid disease. J Immunol 2000; 164:5015-5018.
90. Hill N, Lyons PA, Armitage N et al. NOD Idd5 locus controls insulitis and diabetes and overlaps the orthologous CTLA4/IDDM12 and NRAMP1 locus in humans. Diabetes 2000; 49:1744-1747.
91. Waterhouse P, Penninger JM, Timms E et al. Lymphoproliferative disorders with early lethality in mice deficient in Ctla-4. Science 1995; 270:985-988.
92. Luhder F, Hoglund P, Allison JP et al. Cytotoxic T lymphocyte-associated antigen 4 (CTLA-4) regulates the unfolding of autoimmune diabetes. J Exp Med 1998; 187:427-432.
93. Luhder F, Chambers C, Allison JP et al. Pinpointing when T cell costimulatory receptor CTLA-4 must be engaged to dampen diabetogenic T cells. Proc Natl Acad Sci USA 2000; 97:12204-12209.
94. Salomon B, Lenschow DJ, Rhee L et al. B7/CD28 costimulation is essential for the homeostasis of the CD4+CD25+ immunoregulatory T cells that control autoimmune diabetes. Immunity 2000; 12:431-440.
95. Takahashi T, Tagami T, Yamazaki S et al. Immunologic self-tolerance maintained by CD25+CD4+ regulatory T cells constitutively expressing cytotoxic T lymphocyte-associated antigen 4. J Exp Med 2000; 192:303-309.

Chapter 10

CTLA-4: Its Role in Transplant Tolerance and Rejection

David M. Rothstein and Fadi G. Lakkis

Introduction

Transplantation is the treatment of choice for end-stage heart, kidney, liver, and pancreatic islet disease. Current strategies require life-long immunosuppression in attempts to inhibit the alloimmune response and prevent acute and chronic rejection. Tolerance remains the holy grail for achieving permanent engraftment of transplanted organs while avoiding the attendant risks of chronic immunosuppression such as infection and malignancy.

Allograft rejection is a T cell dependent process. The realization that T cell activation requires signaling through both the T cell receptor for antigen (TCR) and costimulatory molecules, such as CD28, provided new insight into the T cell response. Moreover, the realization that interference with such pathways could alter or block the immune response altogether, provided new strategies for the induction of long-term engraftment through the targeting of T cell signaling molecules. Specifically, interference with the interaction of CD28 with its B7 ligand on antigen presenting cells (APCs) during TCR engagement can induce anergy in T cell clones and prolong allograft survival in a number of animal models. The identification of CTLA-4 as a second ligand for B7 that has potent inhibitory activity in T cells, provided new insight into the nature of tolerance and the balance between positive and negative regulatory signals that regulate T cell activation. Understanding the role of CTLA-4 in allograft rejection and tolerance will assist in designing therapeutic strategies to manipulate positive and negative immunoregulatory pathways.

Negative signals through CTLA-4 are required for the induction and maintenance of peripheral tolerance in a number of experimental models. Mice deficient for CTLA-4 develop a fatal autoimmune disorder that results from dysregulated T cell activation and proliferation in response to environmental antigens.[1,2] Moreover, monoclonal antibody (mAb)-mediated blockade of CTLA-4 prevents development of tolerance to soluble antigen,[3] augments anti-tumor responses,[4] exacerbates autoimmune disease,[5,6] and can even induce de novo autoimmunity in susceptible mouse strains.[7] While crucial in regulating autoimmunity, the role of CTLA-4 in allograft rejection and in the induction of tolerance has been less clear. In regards to the role of CTLA-4 in transplantation, three main questions will be addressed in this chapter: First, what is the role of CTLA-4 in the unmodified allogeneic immune response; second, does CTLA-4 play a critical role in the induction

CTLA-4 in Autoimmune Disease, edited by Flemming Pociot. ©2004 Eurekah.com.

and maintenance of transplantation tolerance; and third, can the CTLA-4 pathway be capitalized upon for therapeutic intervention to achieve tolerance to transplanted organs?

The Role of CTLA-4 in the Allogeneic Immune Response

The blockade of CTLA-4 with anti-CTLA-4 mAbs accelerates and worsen autoimmune diseases including experimental autoimmune encephalomyelitis (EAE) and diabetes in TCR-transgenic mice.[5,6] Moreover, mAb-mediated interference with CTLA-4 can even precipitate spontaneous autoimmune disease in "normal" strains of mice (e.g., BALB/c) that lack apparent propensity towards autoimmunity. Combined with the fatal autoimmune disorder seen in CTLA-4 deficient mice mentioned above, these results provide very strong evidence that CTLA-4 plays a major role in the maintenance of self-tolerance.

In contrast to autoimmune models, short-term treatment of mice with anti-CTLA-4 mAb (three doses) does not accelerate rejection in otherwise untreated murine islet transplant recipients.[8] Moreover, once tolerance is established, three doses of anti-CTLA-4 are unable to induce rejection of cardiac allografts.[9] The autoimmune disease and lymphoproliferation in CTLA-4 deficient mice is driven by antigen-responsive CD4 cells.[10,11] As CD4 cells also play a key role in initiating the alloresponse, it is not immediately clear why CTLA-4 blockade would not hasten rejection and/or break tolerance. One possibility is that not all of the components required for the rejection response are accelerated by CTLA-4 blockade. Thus, CTLA-4 blockade could dysregulate the immune system and predispose to systemic autoimmunity, yet is unable to evoke a specific alloresponse. On the other hand, the apparent inability of anti-CTLA-4 to hasten acute rejection may simply reflect basic differences between auto- and allo- immune responses. Autoimmunity is the result of a chronic exposure to antigen by a relatively small number of autoreactive clones which develop concurrently with host regulatory responses. However, in transplantation, the recipient immune system is acutely exposed to antigen recognized by a large number of alloreactive clones. While immunoregulatory responses may occur, the unmodified allogeneic response is overwhelming and rapid destruction of the allograft ensues. Augmentation of this already aggressive response by anti-CTLA-4 may be difficult to attain or observe. Recent studies support this view. Using a more intensive regimen of six doses of anti-CTLA-4 mAbs to achieve more complete blockade of CTLA-4, the rejection of otherwise untreated recipients of fully MHC mismatched cardiac allografts is indeed shortened in a statistically significant manner – from 7.6 to 5.7 days.[12] However, when the immunologic disparity is decreased, such as in transplantation between mice mismatched at multiple minor histocompatibility loci (e.g., B10-D2 hearts into BALB/C recipients), anti-CTLA-4 had a much more pronounced effect. Here, graft survival was shortened from 96 to 10.6 days. Moreover, CTLA-4 acts on both Th1 and Th2 cells, and pn both CD8 and CD4 cells (see ref.12 and unpublished results). Thus, CTLA-4 plays a role in regulating the allo-immune response that becomes more obvious when the pace of rejection is slowed.

The Role of CTLA-4 Negative-Signaling in Long-Term Engraftment Induced by Tolerogenic Treatment Strategies

Given the acute nature of the alloimmune response, all allograft recipients are treated with immunomodulatory agents from the outset. Targeting T cell activation molecules can alter the immune response and promote long-term graft survival. Specifically, agents that interfere with either T cell activation signal one (through the T cell receptor) or signal two (costimulation), can induce tolerance in mice, even in the face of relatively strong allogeneic barriers. A number of mechanisms including deviation of cytokine secretion, anergy, and apoptosis of alloreactive clones have been shown to contribute to tolerance.

The identification of CTLA-4 as a second ligand for B7 that has potent inhibitory activity provides new insight into the nature of tolerance and the balance between the positive and negative pathways that regulate T cell activation. CTLA-4 plays a key role in maintaining self-tolerance and dampening the immune response in unmodified hosts. However, the role of CTLA-4 in the action of tolerogenic strategies that specifically target T cell signaling pathways was not known until recently. While CTLA-4 might play a global role in the generation of tolerance, it is also possible that agents targeting certain molecules might rely more heavily on CTLA-4-mediated signals. In this regard, CTLA4-Ig is a particularly interesting therapeutic agent. CTLA4-Ig is a fusion protein that combines the B7-binding site of CTLA-4 with an Ig tail. This molecule was designed to take advantage of the fact that CTLA-4 has 10-20 -fold higher affinity for B7 than does CD28.[13] Thus CTLA4-Ig effectively blocks positive signals through B7-CD28 interaction and promotes long-term allograft survival and tolerance in various murine transplant models.[14,15] However, by binding to B7, CTLA4-Ig also has the potential to interfere with CTLA-4-B7 signaling. Thus, the effectiveness of this agent raised the question as to whether blockade of positive signals through CD28 was sufficient to ensure engraftment and what role if any was being played by CTLA-4. To address this issue, we treated murine cardiac allograft recipients with a tolerizing regimen of donor-specific transfusion (DST) and CTLA4-Ig with or without three doses of anti-CTLA-4 mAb in the peri-operative period.[9] Whereas CTLA4-Ig and DST induced long-term engraftment of virtually all recipients, only one of eight mice that also received anti-CTLA-4 maintained their grafts past 100 days (median survival 40 days). Thus, while CTLA4-Ig binds to B7 and blocks its interaction with CD28, preservation of B7/CTLA-4 signaling is required for long term engraftment. Nonetheless, CTLA4-Ig may be somewhat of a double edged sword in that it appears to at least partially interfere with CTLA-4 signaling. Evidence for this comes from studies in CD28 deficient mice. In the absence of dominant positive signals through CD28, CTLA4-Ig treatment actually augments rejection.[16] Thus, CTLA-4 signals are required for engraftment , but CTLA-4 must compete with CTLA4-Ig for B7 binding.

Given its direct interference with regulatory signals through B7, it is easy to comprehend how CTLA4-Ig-mediated graft survival might depend on CTLA-4 signaling. How about the role of CTLA-4 in tolerogenic strategies that act through distinct signaling pathways? In a study by Judge et al, the CTLA-4 dependence of graft survival in mice treated with anti-CD40L mAb plus DST was examined.[9] Although interaction of CD40L with CD40 may upregulate B7 expression on APCs, anti-CD40L is effective in mice lacking CD28,[17] indicating that alternative mechanisms predominate. Interestingly, anti-CTLA-4 did not prevent long-term engraftment induced by anti-CD40L plus DST. In contrast, in a second study, where DST and anti-CD40L were administered in the week prior to transplantation, concomitant administration of anti-CTLA-4 prevented permanent engraftment.[18] Based on these studies, it could be concluded that CTLA-4 signaling has a potent influence on allograft survival, but is not absolutely essential for long-term engraftment when other pathways, such as CD40/CD40L, are targeted.

Based on the premise that tolerogenic agents acting through unrelated pathways might not require CTLA-4 signaling for tolerance induction, we performed studies to determine the role of CTLA-4 in anti-CD45RB-mediated engraftment.[8] Given the critical role of CD45 in TCR-mediated signaling and mAb-mediated alteration in PLCgl phosphorylation,

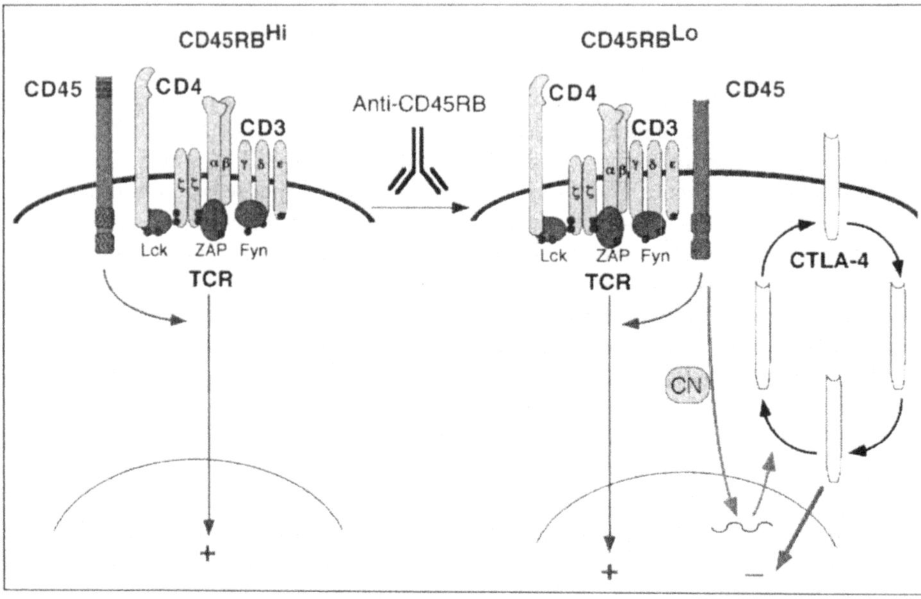

Figure 1. Upregulation of CTLA4 expression by anti-CD45RB. CD45 plays a critical role in T cell activation through the TCR (signal 1). CD4 T cells can be divided into two subsets based upon predominant expression of either high (CD45RB^{Hi}) or low (CD45RB^{Lo}) molecular weight (Mr) isoforms of CD45. Tolerogenic anti-CD45RB mAbs induce a shift in CD45 isoforms expression from high to low Mr. This is not related to depletion or activation of CD45RB^{Hi} cells. The shift in CD45 isoforms is associated with up-regulation of CTLA-4 expression through previously unknown signaling pathways that appear to involve calcineurin (CN). CTLA-4 primarily resides in the cytoplasm, but rapidly cycles to the cell surface upon cellular activation.

anti-CD45RB mAbs are believed to act through alteration of signal one[19-21] (Fig. 1). Whereas anti-CD45RB alone was able to induce long-term islet allograft survival in ~50% of recipients, concomitant treatment with anti-CTLA-4 resulted in acute rejection in all animals by 26 days.[8] The rapidity with which anti-CTLA-4 provoked acute rejection in this model was unexpected and raised the question as to whether CTLA-4 signals might be directly involved in anti-CD45RB-mediated engraftment (discussed further below).

Utilization of the CTLA-4 Pathway for Therapeutic Intervention in Transplantation

CTLA-4 is primarily expressed as an intracellular molecule that cycles to the cell surface where it can then interact with its B7 (CD80 and CD86) counter-ligands APC's.[22,23] CTLA-4 expression is upregulated by T cell activation, normally requiring signals through both the TCR and the CD28 co-stimulatory pathway and entry into cell cycle.[23,24] Unfortunately, soluble agonist ligands for CTLA-4 have not yet been reported, frustrating attempts to directly exploit this downregulatory pathway for therapeutic purposes in transplantation and autoimmunity. As noted above, mAbs against CTLA-4 block its negative signal and thereby augment the immune response-a characteristic useful in promoting anti-tumor immunity but detrimental for allograft survival. Only in the past year have potential approaches for harnessing CTLA-4 for downregulation of the immune response emerged.

Although soluble mAbs block CTLA-4 signaling, negative signals can be induced by CTLA-4 crosslinking and co-localization with activation signals through the TCR. For example, co-localization of anti-CTLA-4 with anti-CD3 and anti-CD28 on plastic plates or latex beads, inhibits T cell proliferation and IL-2 production compared to anti-CD3 and anti-CD28 alone.[25,26] One advantage of using antibodies rather than natural ligand is that CTLA-4 and CD28 can be separately targeted. Although not yet practical, inhibition of in vivo immune responses through Ab-mediated CTLA-4 ligation can be approached through gene therapy. The basis for this resides in a recent study by Griffin et al who expressed a CTLA-4-specific single chain Ab on the surface of APCs.[27] When these APCs were pulsed with peptide Ag, proliferation and cytokine secretion by Ag-specific (TCR-tg) CD4 and CD8 T cells was inhibited. Such inhibition was dependent on co-ligation of CTLA-4 and the TCR/CD3 complex by the same APC. CTLA-4/TCR ligation was inhibitory regardless of whether CD28 costimulation was provided in cis (on the same cell) or trans (on a separate cell). These studies provide evidence that expression of a CTLA-4 ligand by host APCs could downregulate T cell activation and possibly promote tolerance in vivo.

A second means to capitalize on CTLA-4 signaling was revealed in follow-up of studies, alluded to above.[8] The demonstration that anti-CD45RB-mediated engraftment was extremely sensitive to CTLA-4 blockade raised the question as to whether there was a link between CD45 and CTLA-4. In previous studies Metz et al had noted that a subset of "memory" CD4 cells expressing the lower molecular weight (Mr) CD45 isoforms ($CD45RB^{Lo}$) constitutively expressed CTLA-4 (28). Interestingly, we had shown that anti-CD45RB treatment was associated with a rapid shift in CD45 isoform expression from $CD45RB^{Hi}$ to $CD45RB^{Lo}$ [29] (Fig. 1). Putting these data together, we asked whether anti-CD45RB might directly upregulate CTLA-4 expression.[8] Indeed, we found that CTLA-4 was expressed by just 8% of CD4 cells-all expressing the lower Mr isoforms of CD45 ($CD45RB^{Lo}$). Approximately 50% of freshly isolated CD4 cells are normally $CD45RB^{Lo}$. However, treatment with anti-CD45RB results in a rapid shift in isoform expression with >90% of CD4 cells expressing $CD45RB^{Lo}$ isoforms within 1-2 days. This shift in isoform expression was associated with an equally rapid almost two-fold increase in CTLA-4 expression. A time course analysis revealed that CTLA-4 expression on CD4 cells continued to increase such that by day 10 some 25-30% of CD4 cells expressed CTLA-4. Two trivial explanations for the increase in CTLA-4 expression and augmentation of $CD45RB^{Lo}$ cells were ruled out: Anti-CD45RB treatment did not induce overt T cell activation or depletion of $CD45RB^{Hi}$ cells. For example, activation marker expression such as CD69, CD25, and CD44, were unchanged. Moreover, significant depletion of the $CD45RB^{Hi}$ cells did not occur, as there was no detectable decrease in the number of CD4 cells in the spleen or lymph nodes. Furthermore, anti-CD45RB treatment did not alter the bimodal distribution of CD44 (since $CD45RB^{Hi}$ cells normally express low levels of CD44, depletion of $CD45RB^{Hi}$ cells would have been expected to reduce the $CD44^{Lo}$ subset). Thus, mAb-mediated ligation of CD45RB and/or the upregulation of lower Mr CD45 isoforms augments CTLA-4 expression. Moreover, this ensues without overt T cell activation through previously unrecognized pathways.

The biochemical pathways between CTLA-4 and CD45 have not yet been delineated. However this pathway appears to involve calcineurin since treatment with CsA blocks anti-CD45RB mediated upregulation of CTLA-4 and interferes with anti-CD45RB mediated engraftment.[8] These findings have led to the hypothesis that anti-CD45RB may act by rapidly upregulating CTLA-4 expression such that by the time APCs express their costimulatory B7 ligands, potentially alloreactive T cells already express, or have the capacity to rapidly express, CTLA-4 and are subjected to downregulation.

The Role of CTLA-4 and Regulatory Cells in Induction and Maintenance of Engraftment

At present, upregulation of CTLA-4 through anti-CD45RB appears to be the most practical means of harnessing this downregulatory signaling pathway. The degree and length of time that CTLA-4 expression and signaling must be augmented to promote long-term engraftment is not clear. In this regard, the upregulation of CTLA-4 by anti-CD45RB treatment only appears to last several weeks, suggesting that CTLA-4 plays its primary role early, during the induction of tolerance (D.R., unpublished data). Similarly, studies in a cardiac transplant model, suggest that anti-CTLA-4 cannot break tolerance once it has already been established. Whereas, concomitant administration anti-CTLA-4 to allograft recipients treated with CTLA4-Ig plus DST prevented long-term engraftment in seven of eight recipients, delay of anti-CTLA-4 administration by just 8 days results in >100 day engraftment by four of five recipients.[9] Concordant results were obtained in a TCR-transgenic (TCR-tg) model of autoimmune diabetes.[5] In this model, insulitis normally develops within two weeks but diabetes does ensue for five months. However, treatment with three doses of anti-CTLA-4 prior to onset of insulitis induced rapid onset of frank diabetes within two to three weeks. In contrast, when anti-CTLA-4 treatment was delayed until insulitis was already established, anti-CTLA-4 had no effect on either the tempo or penetrance of the disease. Several studies indicate that the mechanisms required to establish tolerance may be distinct from those required to maintain tolerance.[30,31] In particular, maintenance of tolerance may require the presence of regulatory T cells. It is possible that once such regulatory cells are established, or the immune system is skewed towards downregulation, that CTLA-4 no longer plays a requisite role in maintaining allograft survival.

These results are somewhat at odds with the finding anti-CTLA-4 can break self-tolerance and induce spontaneous autoimmunity in otherwise non-autoimmune strains of mice.[7] In this report as well as studies in autoimmune inflammatory bowel disease and diabetes, a regulatory subset of CD25+ CD4 cells was shown to constitutively expresses CTLA-4.[7,32,33] Importantly, the ability of these cells to downregulate autoimmunity may depend on signals through CTLA-4, since anti-CTLA-4 blocked the inhibitory activity of these cells in vitro and triggered autoimmunity in vivo.[7,32] Thus, crucial regulatory subsets of T cells may depend on CTLA-4 signaling for their downregulatory activity. Whether such regulatory cells play a role in the induction of tolerance in transplantation is unknown. It is conceivable that the inhibition of engraftment by anti-CTLA-4 is secondary to inhibition of such regulatory cells rather than blockade of inhibitory signals in allo-responsive T cells. It is also is unknown whether the regulatory CD4 cells that maintain and adoptively transfer tolerance in transplant models either express or depend on CTLA-4.[30,34-37] However, the inability of anti-CTLA-4 to break established allograft tolerance does not currently support a role for CTLA-4 bearing regulatory cells in the maintenance of tolerance.[9] This may change once the identity and mechanism of action of these regulatory cells is better understood. Further work in this area in both transplant and autoimmune models may reveal new ways to manipulation CTLA-4 to our therapeutic advantage.

References

1. Tivol E, Borriello F, Schweitzer A et al. Loss of CTLA-4 leads to massive lymphoproliferation and fatal multiorgan tissue destruction, revealing a critical negative regulatory role of CTLA-4. Immunity 1995; 3:541.
2. Waterhouse P, Penninger M, Timms E et al. Lymphoproliferative disorders with early lethality in mice deficient in Ctla-4. Science 1995; 270:985.
3. Perez VL, Van Parijs L, Biuckians A et al Induction of peripheral T cell tolerance in vivo requires CTLA-4 engagement. Immunity. 1997; 6:411.
4. Leach DR, Krummel MF, Allison JP. Enhancement of antitumor immunity by CTLA-4 blockade. Science 1996; 271:1734.
5. Luhder F, Hoglund P, Allison JP et al. Cytotoxic T lymphocyte-associated antigen 4 (CTLA-4) regulates the unfolding of autoimmune diabetes. J Exp Med 1998; 187:427.
6. Karandikar NJ, Vanderlugt CL, Walunas TL et al. Bluestone CTLA-4: a negative regulator of autoimmune disease. J Exp Med. 1996; 184:783.
7. Takahashi T, Tagami T,Yamazaki S et al. Immunologic self-tolerance maintained by CD25(+)CD4(+) regulatory T cells constitutively expressing cytotoxic T lymphocyte-associated antigen 4. J Exp Med 2000; 192:303.
8. Fecteau S, Basadonna G, Freitas A et al. Upregulation of CTLA-4 expression mediates tolerance induced through CD45. Nature Immunology, in press.
9. Judge TA, Wu Z, Zheng XG et al The role of CD80, CD86, and CTLA4 in alloimmune responses and the induction of long-term allograft survival. J Immunol. 1999; 162:1947.
10. Chambers CA, Sullivan TJ, Allison JP. Lymphoproliferation in CTLA-4-deficient mice is mediated by costimulation-dependent activation of $CD4^+$ T cells. Immunity 1997; 7:885.
11. Oosterwegel MA, Mandelbrot DA, Boyd SD et al. The role of CTLA-4 in regulating Th2 differentiation. J Immunol 1999; 163:2634.
12. Najafian N, Sho M, Kishihara K et al. Regulatory Role Of CTLA-4 And Cytokines In Physiologic Termination Of Alloimmune Responses in vivo. Amer J Transpl 2001;in Press, (Abstract).
13. Linsley PS, Greene JL, Brady W et al. Human B7-1 (CD80) and B7-2 (CD86) bind with similar avidities but distinct kinetics to CD28 and CTLA-4 receptors. Immunity 1994; 1:793.
14. Lin H, Bolling SF, Linsley PS et al. Long-term acceptance of major histocompatibility complex mismatched cardiac allografts induced by CTLA4Ig plus donor-specific transfusion. J Exp Med 1993; 178:1801.
15. Sayegh M, Akalin E, Hancock W et al. CD28-B7 blockade after alloantigenic challenge in vivo inhibits Th1 cytokines but spares Th2. J Exp Med 1995; 181:1869.
16. Lin H, Rathmell JC, Gray GS et al. Cytotoxic T lymphocyte antigen 4 (CTLA4) blockade accelerates the acute rejection of cardiac allografts in CD28-deficient mice: CTLA4 can function independently of CD28. J Exp Med 1998; 188:199.
17. Larsen C, Elwood E, Alexander D et al. Long term acceptance of skin and cardiac allografts after blocking CD40 and CD28 pathways. Nature 1996; 381:434.
18. Zheng XX, Markees TG, Hancock WW et al. CTLA4 signals are required to optimally induce allograft tolerance with combined donor-specific transfusion and anti-CD154 monoclonal antibody treatment. J Immunol 1999; 162:4983.
19. Chan AC, Desai DM, Weiss A. The role of protein tyrosine kinases and protein tyrosine phosphatases in T cell antigen receptor signal transduction. Ann Rev Immunol 1994; 12:555.
20. Koretzky GA, Picus J, Thomas ML et al. Tyrosine phosphatase CD45 is essential for coupling T-cell antigen receptor to the phosphatidyl inositol pathway. Nature 1990; 346:66.
21. Lazarovits A, Poppema S, Zhang Z et al. Prevention and reversal of renal allograft rejection by antibody against CD45RB. Nature 1996; 380:717.
22. Linsley PS, Bradshaw J, Greene J et al. Intracellular trafficking of CTLA-4 and focal localization towards sites of TCR engagement. Immunity 1996; 4:535.
23. Alegre ML, Noel PJ, Eisfelder BJ et al. Regulation of surface and intracellular expression of CTLA4 on mouse T cells. J Immunol 1996; 157:4762.
24. Finn PW, He H, Wang Y et al. Synergistic induction of CTLA-4 expression by costimulation with TCR plus CD28 signals mediated by increased transcription and messenger ribonucleic acid stability. J Immunol 1997; 158:4074.
25. WalunasF/TL, Lenschow DJ, Bakker CY et al. CTLA-4 can function as a negative regulator of T cell activation. Immunity 1994; 1:405.

26. Krummel MF, Allison JP CD28 and CTLA-4 have opposing effects on the response of T cells to stimulation. J Exp Med. 1995; 182:459.
27. Lee KM, Chuang E, Griffin M et al. Molecular basis of T cell inactivation by CTLA-4. Science 1998; 282:2263.
28. Metz DP, Farber DL, Taylor T et al. Differential role of CTLA-4 in regulation of resting memory versus naive CD4 T cell activation. J Immunol 1998; 161:5855.
29. Basadonna G, Auersvald L, Khuong C et al. Antibody mediated targeting of CD45 isoforms: A novel immunotherapeutic strategy. Proc Nat Acad Sci USA 1998; 95:3821.
30. Tran H, Nickerson P, Restifo A et al. Distinct Mechanisms for the induction and maintenance of allograft tolerance with CTL4-Fc Treatment. J Immunol 1997; 159:2232.
31. Lehnert AM, Yi S, Burgess JS et al. Pancreatic islet xenograft tolerance after short-term costimulation blockade is associated with increased CD4+ T cell apoptosis but not immune deviation. Transplantation 2000; 69:1176.
32. Read S, Malmstrom V, Powrie F. Cytotoxic T lymphocyte-associated antigen 4 plays an essential role in the function of CD25(+)CD4(+) regulatory cells that control intestinal inflammation. J Exp Med 2000; 192:295.
33. Salomon B, Lenschow DJ, Rhee L et al. B7/CD28 costimulation is essential for the homeostasis of the CD4+CD25+ immunoregulatory T cells that control autoimmune diabetes. Immunity 2000; 12:431.
34. Qin S, Cobbold S, Pope H et al. "Infectious transplantation tolerance. Science 1993; 259:974.
35. Onodera K, Hancock WW, Graser E et al. Type 2 helper T cell cytokines and the development of "infectious" tolerance in rat cardiac allograft recipients. J Immunol 1997; 158:1572.
36. Gao Z, Zhong R, Jiang J et al. Adoptively transferable tolerance induced by CD45RB monoclonal antibody. J Am Soc Nephrol 1999; 10:374.
37. Bushell A, Niimi M, Morris PJ et al. Evidence for immune regulation in the induction of transplantation tolerance: a conditional but limited role for IL-4. J Immunol 1999; 162:1359.

Index

I

L

M

N

P

R

S

T

For Product Safety Concerns and Information please contact our EU representative GPSR@taylorandfrancis.com Taylor & Francis Verlag GmbH, Kaufingerstraße 24, 80331 München, Germany

Printed and bound by CPI Group (UK) Ltd, Croydon, CR0 4YY
14/09/2026
02216256-0008